Through the Telescope
A Guide for
the Amateur Astronomer

Michael R. Porcellino

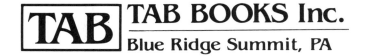

TAB BOOKS Inc.
Blue Ridge Summit, PA

Dedication

I dedicate this book, my first, to three people who inspired me more than they ever knew or could know:

To my mother and father, Angela and Robert Porcellino, who started me on my way by giving me my first telescope when I was 12. Thanks, I only wish you both could be here to see this.

And to my lovely wife Barbara because she is....

FIRST EDITION
FIRST PRINTING

Library of Congress Cataloging-in-Publication Data

Porcellino, Michael R.
    Through the telescope : a guide for the amateur astronomer / by
  Michael R. Porcellino.
      p. cm.
    Bibliography: p.
    Includes index.
    ISBN 0-8306-1459-1   ISBN 0-8306-3159-3 (pbk.)
    1. Astronomy—Amateurs' manuals. 2. Astronomy—Observers'
  manuals. I. Title.
  QB63.P59 1989
  522'.2—dc20   89-34641
                                       CIP

TAB BOOKS Inc. offers software for sale. For information and a catalog, please contact TAB Software Department, Blue Ridge Summit, PA 17294-0850.

Questions regarding the content of this book should be addressed to:

    Reader Inquiry Branch
    TAB BOOKS Inc.
    Blue Ridge Summit, PA 17294-0214

Acquisitions Editor: Roland S. Phelps
Technical Editor: Barbara B. Minich
Production: Katherine Brown

Cover photograph by Walter Piorkowski, Chicago, IL.

Contents

Acknowledgments

A book is never the work of only one person. By meeting and talking to a great number of people, an author is able to garner bits of information that, in one form or another, eventually make their way into his work. Over the course of writing this book, I have had the pleasure and the honor to meet, talk, and correspond with amateurs from around the country. In their way, however small, many people have contributed to the completion of this book.

I would like to thank Jim Seevers and the staff of the Adler Planetarium of Chicago for allowing me to use their excellent library during the research of this book. I would especially like to thank librarians Pamela Hovan and Evelyn Natividad for putting up with me while I rummaged through books at the back table.

I would like to thank Jeffrey Hunt of Cosmic Connections for reading most of this manuscript and making invaluable comments about its content.

The members of the Chicago Astronomical Society, whether they know it or not, helped with many aspects of the work by their comments. I would especially like to thank Jim Carroll, John Jones, Robert Stalzer, and Kathy Piorkowski for their contributions and suggestions.

Other members of the amateur astronomical community from around the country also helped with this project. I want to thank Chet Beck, Paul Bock, Robert Browning, Tom Johnson, and Jim Senk for their letters and comments that helped develop the book from concept to completion.

Thanks also to Pat Shand and the staff of Lick Observatory for some of the photographs found in this book. Jeff Beish, Don Parker, Jim Phillips, and the other members of the Association of Lunar and Planetary Observers (ALPO) helped, as did Janet Mattei and the American Association of Variable Star Observers (AAVSO). I thank you all.

I want to extend a special thanks to Dave and Marion Bachtell, Barbara Lux, Dan Joyce, Lee Keith, Walter Piorkowski, Dan Troiani, Suresh Sreenivasan, Mark Stauffer, and Linda Warren for the drawings, photographs, and ideas they contributed to this book. Without their help it would never have been completed.

Any errors or omissions found in this work are mine and mine alone.

Finally, I want to thank Roland Phelps, my editor at TAB BOOKS, for putting up with me during the course of this project.

Introduction

Welcome to the wonderful universe of amateur astronomy! Amateur astronomy is something that anyone can enjoy. All that is needed are clear skies and a dark place to start you on the road to the stars. While no one can supply clear skies, this book can start you on your way.

I wrote this book with two kinds of amateurs in mind. First, it is written for the beginner who is thinking of getting, or has just gotten, a telescope and wants to learn just what it can show them. If you find yourself in this group, you have to take some time and prepare yourself. A big mistake that many people make is to rush out and get a telescope before they are ready to use it.

When is a person ready for a telescope? You must have a working knowledge of the constellations and the star figures that you are going to wander through with your telescope. You wouldn't leave home for a long trip without some idea of where you were going, would you? Well, that's just what you are doing without a knowledge of the constellations. If you don't know where Epsilon Bootes is, then you are never going to be able to find the variable star W Bootes. So take some time and familiarize yourself with the stars. When you know your way around the stars that you can see with your naked eye, then you are ready for a telescope and the stars that you cannot see!

The other group I wrote this book for includes those with a small telescope. They have reached a point where they can easily find their way around the skies, and they have invested in a telescope but they don't know where to point it. This book will serve as a guide to using that instrument. Many of today's observing guides tend to pass over these amateurs, an omission I find unfortunate, and these small scope users feel they have been cheated.

Unfortunately, in this age of 17.5-inch dobsonians, too many of today's amateurs look down on the under 4-inch telescope like a poor relative. But if you ask 100 amateur astronomers how big their first telescope was, I'll wager that 85 percent of them will say it was less than 4 inches. These amateurs and the writers of observation guides have forgotten their "roots." Since that first telescope they have moved up to the huge "light buckets" that have become all the rage, and they now have little use for the small telescope. Everyone in amateur astronomy should remember that they started with some kind of telescope. In many cases it was a small one.

To meet the needs of both these segments of the astronomical community, I wrote this book as a survey of the wide variety of experiences that an amateur with a telescope can enjoy. I've tried to give an overview of the equipment needed by the observer and some thoughts on its selection. I've also tried to show that an amateur with a small telescope can make a contribution to science. Sure, some activities require a large-aperture instrument, like supernova searches, but other areas just as important to science require only a small instrument.

Whether you are just entering the wonderful universe of amateur astronomy or have been in it for awhile, I want you to remember one thing above all else. Amateur astronomy should be fun! This is not a contest to see who can spot the most features on the planet Mars or estimate 100 variables in one night. It is an immensely enjoyable, relaxing way to spend your time, and if you can make some contribution to science along the way that's great, but it is not a requirement.

So enjoy the stars, enjoy your telescope, and most of all enjoy this book!

Clear skies to you all!

✳1✳

The Amateur
In Astronomy

*Astronomy offers a pleasure that follows the law of
increasing rather than diminishing returns.
The more you develop a thirst for it, the greater your return.*

William T. Olcott

There is something almost mystical about sitting under a crystal clear night sky while the strains of Bach or the Beatles play on your portable stereo. Sitting and studying the shapes of the constellations as they wheel across the black velvet dome of the sky is both exciting and relaxing. This is the realm of the amateur astronomer.

There are many definitions of the word *amateur*. A few listed by Webster's Dictionary include "dilettante," "nonprofessional," and my personal favorite, "one who is unskilled." These definitions might be adequate for an entry in Webster's, but they hardly describe the amateur in amateur astronomer. These definitions forget that *amateur* comes from the Latin word *amator*, which means "one who loves." That describes the amateur astronomer; it describes one who loves astronomy.

THE AMATEUR TRADITION IN ASTRONOMY

During most of the eighteenth century, astronomy was a practical science. The job of the astronomer was simple: produce information that would be useful in the exploration of the world. So the astronomer tried to determine the positions of stars as accurately as possible. This kind of information was just the kind of thing a sailor or surveyor needed to know so that he could figure out where he was.

Until the late eighteenth century, positional astronomy was the order of the day at the observatory. In many observatories it was forbidden for the staff to use the tele-

1

scopes for anything but the measurement of star positions. It was also a time when many who held the title astronomer had to work their way up the ladder from the boring jobs of calculating ephemerides and copying charts.

This situation began to change in the mid-1700s with the invention of the achromatic lens system for the refracting telescopes by John Dolland. Apprentice astronomers, taken with the images of the stars during their stints at the transit telescope, began to get curious. What would the Moon look like with this instrument? They would risk the wrath of their employers and sneak a peek. Soon, even their employers were peeking into the universe. Others began experimenting by casting metal mirrors and constructing instruments along the lines suggested by Sir Isaac Newton. Private affairs, these telescopes were required to do nothing more than show their owners the universe.

Fig. 1-1. Sir William Herschel is considered to be the founder of observational astronomy. This diorama at Chicago's Adler Planetarium shows what observing was probably like in the late 1700s. The telescope in the picture is a 6.2-inch reflector made by Herschel. (Courtesy Adler Planetarium.)

The man credited with the invention of observational astronomy, Sir William Herschel (Fig. 1-1), was by trade a musician. Each night he would retire with a glass of milk and two books: one on harmonics, the study of musical sounds, and the other on astronomy. His interest in astronomy piqued, he began constructing simple telescopes from spectacle lenses. In 1774 he began constructing large reflecting telescopes and using them for a systematic study of the sky for no other purpose than to see what was out there. Observational astronomy had begun.

In the United States, no one exemplified the amateur tradition in astronomy better than Sherburne Wesley Burnham. Born in Vermont in 1838, Burnham was trained as a shorthand writer. He served in the Civil War as a court reporter with the Union forces occupying New Orleans, and later as the Chief Clerk of the Circuit Court in Chicago. He would work eight hours at his job and then stay up far into the night studying double stars from the backyard of his home in Chicago. Using a 6-inch refractor built for him by the Massachusetts firm of Alvan Clark & Sons, Burnham discovered and studied hundreds of double stars. His publication of his observations lead to recognition by scientific societies around the world and entry into some of the world's most prestigious observatories. With the exception of a two-year stint on the staff of Lick Observatory, all of Burnham's work, until his death in 1921, was done as an amateur.

In the twentieth century, the amateur tradition has been represented by a number of observers. Probably best known among them was Leslie Peltier of Delphos, Ohio. Peltier began his love of the heavens as a boy and carried the torch until his death in 1980. Called the "world's best nonprofessional astronomer" by Harvard College Observatory director Harlow Shapley, Peltier's career spanned almost 70 years. During this time he discovered 12 comets, contributed over 100,000 brightness estimates of variable stars, and made numerous other accomplishments in the field of astronomy.

Today, the tradition continues. A doctor in South Carolina spends his time away from the hospital studying features on the Moon known as domes in an attempt to understand lunar volcanic action. A postal worker in Chicago rises well before dawn to study Mars and contribute to our growing knowledge of that planet's weather. A chemist in Baton Rouge meticulously measures the distance between double stars in an attempt to refine the knowledge gained about them over the past 200 years.

WHY AMATEUR ASTRONOMY?

People become interested in astronomy for a variety of reasons. Ask 100 amateur astronomers how they became interested and you will probably get 100 different answers. Many will admit that using a small telescope they received as a child began a life long love affair with the stars. For others it was a trip to a planetarium or an event sponsored by a local astronomy club that sparked their interest (Fig. 1-2).

Some were simply fascinated by the patterns the stars created in the night sky. After spending some time in the local library reading up on stars, constellations, and astronomy they soon found themselves able to identify such star groups as Orion, Lyra, and Leo. Still others owe their interest in astronomy to their children. Helping with a science project has given many a parent a taste for science in general and astronomy in particular.

Regardless of how your interest in astronomy was sparked, there are a number of ways you can get more involved with astronomy. First you can simply experience the sky

Fig. 1-2. People from all over the country attend Chicago Astronomical Society's annual Astro-Fest.

with a fascination that has held mankind since the days of the cave. If there is a pair of binoculars around the house, you can take it out and turn it toward the stars. Once you have a taste of what lies beyond the grasp of the naked eye, you will realize that your appetite has only been whetted. Finally, you might buy or make a telescope. Each of these stages is a learning experience and if progressed through properly, you may eventually make significant contributions to science.

You could soon find yourself rising at 3:00 A.M. to get a good view of Mars or Jupiter. You might find yourself thrilled by the changing brightness of a distant star and record those changes at every opportunity. The rolling line of sunrise on the Moon may catch your fancy, and you might soon find yourself watching it night after night until you realize that the distant surface of the Moon is as familiar to you as your backyard.

CONTRIBUTIONS FROM THE AMATEUR

Amateur astronomers fill a void in today's science. Unfortunately, budget restrictions at many of the major observatories make some types of astronomical work uneconomical.

The amateur is relied upon to fill the void. A prime example is the study of the planets. With the *Viking* missions to Mars and the *Voyager* missions to Jupiter and Saturn many astronomers felt that the days of Earth-based planetary observation were over.

These missions however, despite their mind-boggling results, last only a finite time. If we are going to launch and carry out a manned mission to Mars, for example, we have to have a better understanding of Martian weather. It doesn't do any good to mount an expensive manned mission that will last two or more years and travel over 120 million miles only to have the spacecraft orbit the Red Planet unable to land due to a massive dust storm. Our knowledge of Martian weather and the growth and predictions of major dust storms have come from the work of amateur astronomers like Donald Parker, Daniel Troiani, and other members of the Association of Lunar and Planetary Observers (ALPO).

The internal constitutions of the stars, long the purview of professional astronomers, have also felt the impact of the amateur astronomer. Now that observational time on large telescopes is severely limited, amateurs are contributing observations of the fluctuations in brightness of stars. This information helps scientists determine what makes a star shine. Since its foundation in 1911, the members of the American Association of Variable Star Observers (AAVSO) have contributed over 5 million observations of stellar brightness. The role played by the members of AAVSO has also grown in importance with the introduction of sophisticated satellites that study the stars from Earth's orbit. Often amateur observations provided by AAVSO enable scientists to correlate an event observed by their satellites with actual visual observations.

Don't get me wrong, there is nothing that says you must, as an amateur astronomer, make such observations. The beauty of amateur astronomy is that you can make contributions like these if you want to, not because you must. Remember the root word for amateur is amator meaning one who loves astronomy.

ONE AMATEUR'S EXPERIENCE

Just what kind of contribution can a single amateur make to the world of astronomy? At first glance it would seem hopeless to try to compete with the huge telescopes and complex instruments of the professional, but have faith, it is possible.

Dan Troiani first became interested in astronomy in 1974 when he visited Chicago's Adler Planetarium with his future wife, Kathy. He was fascinated with the sky show and the wide variety of astronomical exhibits at Adler. He bean to read about astronomy and became more deeply interested. He also became an avid Star Trek fan, but found he was more interested in the "science" than the "fiction." "I realized after a short time," he said, "that the science of astronomy was stranger and weirder than fiction could ever be."

In his wide readings, Dan was taken by the amount of detail visible on the planetary drawings he saw. "Could I do that?" he asked himself. The skies over a large city like Chicago are not conducive to many types of astronomical work. It became apparent to Dan during the course of his research that the Moon and planets were the perfect objects for a city astronomer to study.

In the 1970s, before the revolution in telescope design brought about by lightweight mirrors and the dobsonian mount, a 10- or 12.5-inch newtonian reflector was considered a large telescope. Dan decided that a 10-inch reflector would be the perfect tool for a serious program of planetary study. So in March of 1977 he ordered a 10-inch mirror from Cave Optical Company and constructed the telescope in Fig. 1-3. He also joined the

Fig. 1-3. *Daniel Troiani of Chicago, Illinois and the 10-inch reflector with which he rediscovered the Martian polar feature Rima Tenuis.*

Chicago Astronomical Society and learned how to use his telescope. After a short construction period, Dan began an observational program that concentrated on the bright planets. He joined ALPO in late-1978 and began to contribute his observations to that group. In 1979 he widened his astronomical program, joined AAVSO, and began to study variable stars. Everything went smoothly for Dan until December 17, 1979.

We have to stop here and drop back in time almost 100 years. During the 1880s the famous Mars observer Giovanni Schiaparelli, director of the Brera Observatory in Milan, Italy, noted that the north polar cap of Mars seemed to be divided into two parts. He determined that the cause of the division was a dark rift in the polar cap. He called this rift Rima Tenuis.

The rift was repeatedly observed during the next close approaches of Mars until the approach of 1918. The Rima Tenuis had disappeared! It was searched for repeatedly during following years but could not be seen. During the 1960s, Rima Tenuis was searched for using telescopes from 24- to 82-inches in diameter but these large instruments found nothing. When the *Mariner* and later *Viking* spacecraft arrived at the planet, their photography showed nothing at the north polar cap but layers of white. Rima Tenuis had indeed vanished.

On December 17, 1979, Dan was at his 10-inch telescope making drawings of Mars to forward to ALPO for study. He noted a dark rift that seemed to cut across the north polar cap and recorded it in his drawing (Fig. 1-4). At the time Dan did not think much about the significance of what he was seeing. ''I just tried to get it down accurately in the drawing,'' he said.

When Dan's drawing was received by ALPO's Mars director a few days later, it was looked at critically and with some disbelief. ''I guess they didn't trust my eyesight,'' Dan says with a shrug. Astronomers around the world in ALPO's network of Mars observers rushed to their telescopes to see if Dan had really seen something that had not been

Fig. 1-4. Dan Troiani's drawing of the Rima Tenuis. (Courtesy Dan Troiani.)

visible on the Red Planet for almost 60 years. But for six weeks they had to wait. Because of the way Mars rotates, features visible near the edge of the planet pass behind Mars and are out of sight for that long. Finally, in February 1980 confirmations began to arrive: Rima Tenuis was back! Dan Troiani had succeeded with a 10-inch telescope where instruments two to eight times larger had failed. He had seen more from his backyard in Chicago than advanced spacecraft had seen from an orbit a few hundred miles above the planet. The credit for the rediscovery of Rima Tenuis on the north polar cap belongs to Dan Troiani, amateur astronomer, and no one else!

See! I told you it was possible!

✳2✳

The Eye and Astronomy

. . . fortunate is he whose introduction to the skies comes
. . . through nature's eyes alone and not through any telescope.

Leslie Peltier

The basic instrument of the astronomer is the eye. Until Galileo Galilei came along in 1609 and turned his *optik tube* on the heavens, astronomy was done with the human eye. Early astronomers used their visual skills to lay the groundwork that has taken us to our present level of knowledge. Greek astronomers were able to explain the phases of the Moon, devise a theory of planetary motions that Nicolaus Copernicus later revised and improved, measure the distance to the Sun and the Moon, and determine the shape and size of the Earth.

Tycho Brahe, a Danish nobleman and astronomer, compiled accurate observations of planetary positions using some simple sighting instruments. His assistant, Johannes Kepler, used this mass of data as the basis for his three laws of motion. Without the contributions of these diligent observers, Sir Isaac Newton would have had no shoulders to stand on and the history of science might have been very different.

All of this work was accomplished using nothing but the eye. Anyone who wishes can apply the same techniques and can see that they yield the same results. Yet one of the hardest things for the beginning or casual observer to understand is how you actually see and how this relates to astronomy. Whether you observe with the naked eye, a pair of 10×50 binoculars or an 8-inch telescope, everything you see has to pass through the eye.

HOW THE EYE WORKS

When you look at a star, its image enters the eye by passing through the *cornea* (Fig. 2-1) where the image is focused. The brightness of the image is controlled by the varying diameter of the *pupil*, which opens or closes depending on the amount of light entering it. When you are looking at a bright object, like the Moon, your pupils will *contract*, or close, because a large amount of light is reaching the eye. When you are looking at a very faint object, like a galaxy or a very faint variable, the pupil will *dilate*, or open, to allow more light to reach the eye (Fig. 2-2). The average size of the exposed pupil varies throughout life, but averages between 6mm and 7mm. At 20 years of age, the pupil is in its best health and can be as large as 8mm when fully opened in low light. As a person ages, the effectiveness of the pupil diminishes. At age 50, the pupil can only open to 5mm, and by age 80 the pupil can expand only to a diameter of 2mm.

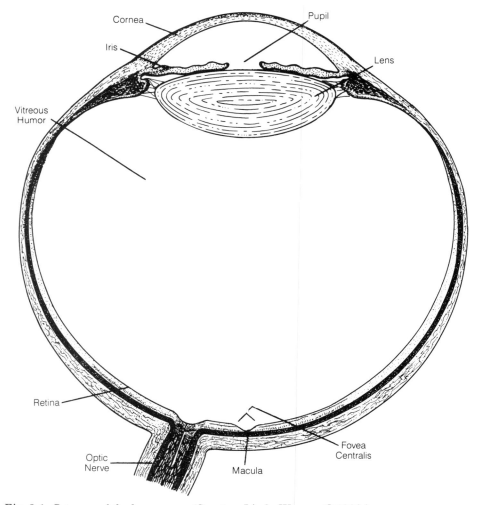

Fig. 2-1. Cutaway of the human eye. (Courtesy Linda Warren, ©1988.)

Having passed through the cornea and the pupil, the image is given a fine focus by the *lens*. After moving through the central globe of the eye, the image falls on the *retina*. It is here that the work is done.

The real power of the eye is located in the retina (Fig. 2-3). The retina is a thin layer of cells that line the back of the eye. It is the only part of the central nervous system that is exposed to the outside world and direct stimuli. The retina is composed of five types of cells found in four layers.

The retina is able to change its sensitivity level and adjust to incoming visual signals. It is able, in effect, to be selective in what it passes on to the brain. The retina can turn areas of itself on or off as it needs them to analyze the image using its outer two cells layers.

Located behind these two layers are the *photoreceptors*. They are anchored to a layer of cells from which they grow and draw nourishment. Because of the way they are attached, the photoreceptors face away from the incoming image, which must pass through them before triggering a response.

There are two kinds of photoreceptors. The kind that give us our ability to see color are called *cones*. An individual cone can be sensitive to one of three primary colors: blue, green, or yellow. The mixture of the signals from these three primary color receptors gives us color vision.

The concentration of cones moves outward from an area of the retina called the *fovea centralis* (Fig. 2-4), which contains only cones, through the central part of the retina, called the *macula*. A few degrees out from the fovea centralis the cones lose their dominance to the second kind of photoreceptor, called *rods*.

Outnumbering the cones 20 to 1, rods are more sensitive to low light levels than cones. The rods play no roll in color vision. Their primary purpose is to act as light collectors. Except within the central area of the retina, rods dominate. The tremendous

Fig. 2-2. The pupil and iris act together to admit light to the eye. The diameter of the fully open pupil is between 6 and 7mm. (Courtesy Linda Warren, ©1988.)

number of rods allows the retina to gather enough light to see under almost any circumstances.

Because of the range of light intensity with which the eye must function, it has developed a dual system for vision. One system is designed to see fine detail and detect

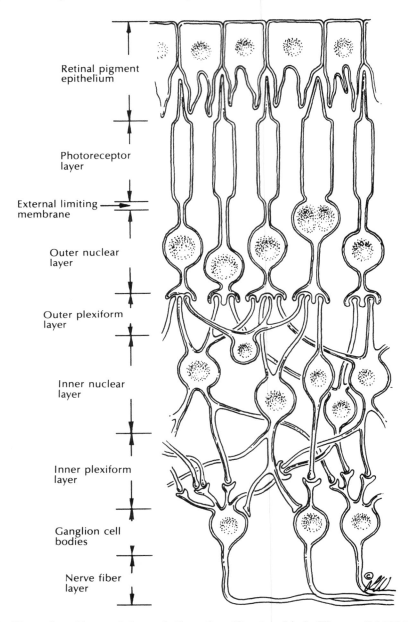

Retinal pigment epithelium

Photoreceptor layer

External limiting membrane

Outer nuclear layer

Outer plexiform layer

Inner nuclear layer

Inner plexiform layer

Ganglion cell bodies

Nerve fiber layer

Fig. 2-3. The real workhorse of the eye is the retina. (Courtesy Linda Warren, ©1988.)

rapid movement. This system, called *photopic vision*, makes use of the 6 million cones concentrated in the central area of the eye. In the fovea centralis, vision is the sharpest and the eye has its highest resolving power. In order for this system to work, however, light levels must be extremely high. This is the system of vision that allows you to read the letters printed on this page.

The other system, *scotopic vision*, functions when the light is dim, such as when you are looking through a telescope at a faint galaxy. Low light requires the eye to utilize the 130 million rods that are found everywhere in the eye except in the fovea centralis. These rods gather light from a large area over a longer period of time. You may be able to see the white of this page if you flip off the light. As your eyes adapt to the dark, you may even be able to distinguish the letters on the page as so many smudges, but you will not be able to read the words using this system of vision.

DARK ADAPTATION

The ability of the eye to adapt to low light levels is very important to the observer. With the exception of a few objects like the Moon and some of the planets, most light levels encountered by the amateur astronomer are very low. Adequate dark adaptation allows you to see details in galaxies and nebulae that the unadapted eye would let slip by unobserved.

When you expose your eye to low light levels, a number of physical changes take place. First, the pupil opens to allow more light to enter and strike the retina. Despite the greater amount of light entering the eye, only about half of it actually reaches the retina. The rest is absorbed by the fluids in the eye.

A chemical produced by the retina called *rhodopsin*, or visual purple, bathes the rods and helps the cells make the most of the small amount of light now reaching them. The chemically charged rods interact with the other cells to enhance the image before it is passed on to the brain. You can see this happen by sitting in a darkened room. At first you can see little, if anything. The longer you remain in the low light, however, the more the details around you become apparent.

Fig. 2-4. The cones, the photoreceptors responsible for our color vision, are grouped in the fovea centralis. Surrounding the fovea are the millions of rods that enable us to see low light.

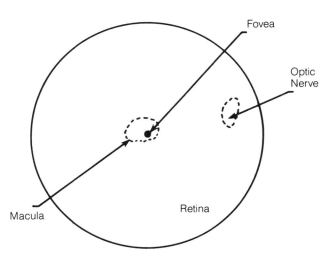

You need to encourage those same changes to take place before you can see minute detail with the telescope. The entire dark adaptation process takes about 30 minutes. After the first minute of dark adaptation, the eye is 10 times more sensitive to light. After 20 minutes it is 6,000 times more sensitive. After 30 minutes, the eye reaches its limit and the fully adapted eye now requires 30,000 times less light than normal to activate the retina.

William Herschel, considered to be the founder of observational astronomy, went to extremes to obtain what he called "tranquility of the retina." He would spend 20 to 30 minutes in complete darkness before he would try to observe faint objects. His dark adaptation was so complete that he would be forced to stop observing if a third-magnitude star came into his field of view. The light would almost blind him! Although it is not necessary for you to sit in a darkened room, you can pass the time needed to adapt your eyes by sitting out under the sky and studying the constellations.

THE LIMITS OF VISION

The ability of any optical instrument is measured against the eye. The limits of the eye act as a baseline for comparison. It is important to understand these limits so we can learn how to make astronomical instruments that exceed them.

The primary thing the astronomer has to contend with is the fact that the eye can take in only so much light. The light gathering power of any optical system is determined by the size of the area able to receive light. In the eye, this is the diameter of the fully dilated pupil. In a telescope it is the diameter of the lens or mirror. The amount of light that the eye, or optical surface, is able to gather in determines how faint an object it can detect.

The brightness or dimness of stars and other celestial objects is measured in *magnitude*. One of the first things you notice when you look at the stars is that they are not all the same brightness. Some are plainly visible even through the light pollution of the city. Others are only visible well away from city lights.

The magnitude system was devised by the Greek astronomer Hipparchus. Virtually unchanged since 150 B.C., Sirius, Castor, Pollux, Vega, Deneb, and the other bright stars were lumped together as first-magnitude stars. Slightly fainter stars became the second magnitude. Hipparchus continued to categorize all the stars he could see. The final group in his system, stars of the sixth magnitude, represent the limit of human vision.

Science has improved slightly on Hipparchus' system. Complex photometric devices instead of the naked eye, now measure the brightness of the stars. The range of the magnitude scale has been vastly extended. Today, the dimmest star visible to the largest telescope is 25th magnitude. But the basic ratio of brightness remains unchanged. Each magnitude is 2.5 times brighter than the next magnitude. A first-magnitude star is 2.5 times brighter than a second-magnitude star and 6.25 times (2.5×2.5) brighter than a third-magnitude star. Sixth-magnitude stars are 100 times fainter than stars of the first magnitude.

Remember, however, that the limit of human vision is dependent on a number of outside factors. Your *limiting magnitude*, the faintest star you can see, will be affected by your observing site. In the city it is difficult to see down to the sixth magnitude; there are times when you can barely make out first-magnitude stars. On most nights in Chicago I can usually see to fourth magnitude. Someone with exceptional eyesight may be able to see to seventh magnitude under clear country skies.

If you hold this book 12 inches from your eyes, you can plainly see each letter on the page. The eye is able to separate each letter into a distinct and recognizable shape; it can resolve the letters. If you move the book back 5 feet, can you still distinguish the letters? How about at 10 feet? At 20 feet? Somewhere down the line the eye stops seeing individual letters and they all run together. At that point, the eye has reached the limit of its *resolution*.

In astronomy, size and distance are measured in degrees. Each degree is made up of 60 arcminutes, and each arcminute is further divided into 60 arcseconds. Resolving power is measured with these smaller units. You can easily see the full disk of the Moon because it is ½ degree, or 30 arcminutes, in diameter. The star at the bend in the handle of the Big Dipper has a companion at a distance of about 12 arcminutes. The ability to see Mizar and its companion Alcor was a test of visual acuity in many cultures. The generally accepted limit for the unaided eye is about 4 arcminutes, or 240 arcseconds. The double star Epsilon Lyrae is a good test of visual acuity. With its components 208 arcseconds apart, it is just below the threshold of the unaided eye. A person with very good vision should be just able to separate the pair.

LIGHT POLLUTION

The ability of your eyes to see anything in the night sky is contingent on one thing: that you can actually see stars. Amateurs in the major cities of the United States and around the world are faced with a growing menace: light pollution (Fig. 2-5). Street lights installed to curtail crime and make the streets safe have created a nightmare for many urban observers. Amateurs living in rural areas are not unaffected by this growing problem. Ten years ago an amateur living 20 miles outside a major city like Chicago had a beautiful view of the sky. His observing is now tainted with the spreading glow of the city's lights, to say nothing of the lighting "improvements" installed by the suburb.

Light pollution problems fall into one of the three general categories. *Light trespass* is simply light spilling from one location onto another. If your neighbor has a security light that is on all night and interferes with your observing, you are the victim of light trespass. Light trespass deals a death blow to the dark adaptation you have worked so hard to achieve and eliminates many of the faint, fuzzy objects that inhabit the night sky. The best way to deal with this is to diplomatically ask your neighbor to turn off the light. Offer him a look through your telescope and in 99.99 percent of the cases he will be happy to oblige. If the offending light is a streetlight placed by the city, it is not a lost cause. Try to position yourself so that the offending light is blocked by a building or trees, or erect some kind of shelter to block the light.

Working hand-in-hand with light trespass is *glare*. We have come to expect a good light to be both dazzling and omnipresent. A good light, however, should not blind you when you look toward it.

The final category is *urban sky glow*. This has lead to major problems for large observatories like Mount Palomar in California, the Kitt Peak National Observatory in Arizona, and the U.S. Naval Observatory in Washington, D.C. The thought of having to

Fig. 2-5. Light pollution is a threat to all astronomers. This satellite photo gives you some idea of the severity of the problem. (Courtesy USAF.)

Fig. 2-6. Many college campuses are switching to more effective, shielded lighting units, which direct the light at the ground where it is most needed.

put these magnificent instruments in dark storage because of the encroaching lights of urban progress has rallied many professional astronomers to action.

The obvious culprits in this battle against light pollution are lights. Now don't get me wrong, professional and amateur astronomers are not against nighttime lighting. Like everything else, however, there is a right way and a wrong way to illuminate the night. It makes no sense to use lights that waste more than 30% of their effectiveness by spilling light up into the sky. Light that goes straight up is not useful in cutting down crime or protecting the citizens of the city. Pure and simple, it is light that is wasted! "About the only useful purpose for lights like this," says Dr. David Crawford of the Kitt Peak National Observatory, "is to light up the underbelly of airplanes." It is simple to install lights and fixtures that put light where it is most needed—on the ground (Fig. 2-7).

Most cities and industries today use what is called high-pressure sodium (HPS) light. This type of light is not energy efficient, nor are the fixtures in which these lights are mounted. A better solution would be to use a more economical type of light and shield it to keep the excess light from seeping into the sky. Does this type of light exist? Yes. It is called low-pressure sodium (LPS). When used in conjunction with properly shielded lighting fixtures it reduces light pollution and improves the overall economy of the lighting system.

Another plus for LPS lighting is the fact that the light emitted by these fixtures is restricted to a narrow band of yellow light that is easily filtered. LPS lights are easy to recognize by their yellow glow. HPS lights put out their light over a much broader spectrum, and are almost impossible to filter out. They are identified by their pinkish color.

Agreements between astronomers and the city of Tucson, Arizona, has lead to the installation of the more efficient LPS lighting system. This saves the city money and allows the nearby observatory to filter the light for more efficient viewing. It is a solution that lets everyone win.

While professionals have faired rather well in the fight against light pollution, amateurs have not been so lucky. The Milwaukee Astronomical Society's observatory complex in New Berlin, Wisconsin, has been fighting urban expansion and its accompanying light pollution for years. They approached the New Berlin city council with a plan to protect the night skies over their observatory. Although the city was sympathetic to the pleas of the amateurs, the council turned them down. The observatory complex is now in danger of losing its sky to light pollution.

What can the amateur astronomer do about light pollution? The entire issue comes down to the simple task of education. Many people outside of astronomy don't know the first thing about light pollution, and you, as an affected observer, can help introduce them to the problems and solutions concerning light pollution. You can show people, especially neighbors who have those wonderful security lights, how to improve them so they are more effective and efficient. If you are successful, your neighbor can get a lower energy bill and you can get a darker observing site. Again, everybody wins!

The International Dark-Sky Association (IDA) has been organized for the purpose of preserving dark skies for astronomy. Dr. David Crawford, one of IDA's founders, estimates that in a few generations "the majority of people in the world will never have seen the universe. The only place they will be able to see it is in a planetarium or on a TV screen. And that," he adds, "is sick!"

NAKED-EYE ASTRONOMY

Probably the most enjoyable way to discover the universe is with the eye alone. A large number of objects and phenomena can be observed using nothing more than a comfortable chair and your eyes.

The first thing to do if you are going to enter the wonderful world of amateur astronomy is to learn the constellations. Now this may sound like a real task but believe me, it's fun! Learning the shapes and stories behind the star figures that move through the night sky can become an avocation in itself.

I've heard a lot of people complain that they just can't see much from the city. Living in Chicago all of my life, I have seen the skies gradually deteriorate, so I can understand what they are saying. I can remember the skies of Chicago I saw as a boy when if you looked hard enough you could just make out the Milky Way. Not anymore! From where I observe, within a mile or so of Midway Airport and right under a landing path, I am happy when I can make out stars of the fourth magnitude!

But strange as it may seem, learning the sky under conditions like this can be a blessing in disguise. The sky as seen from a dark site is so strewn with stars that it can confuse new observers. Such confusion can lead to a decision to find another outlet for their scientific curiosity. Learning under the limited skies of a city like Chicago allows you to learn the brightest stars in each constellation and makes the transition to darker skies and more stars easier.

I've listed a few of the many constellation guides available to you in the Bibliography at the end of this book. My advice is for you to go to your local library and study one or two of them. Take them home and leaf through them at your leisure. Remember you will not be tested on any of this, so there is no pressure. Take the time to learn the constellations. If you never venture any further into the world of amateur astronomy, you will have a treasury of knowledge that no one can take away from you.

✳3✳

Binocular
Astronomy

*. . . the stars . . . appear with exquisite and startling beauty
through a field glass.*

William T. Olcott

The eye may be the basic instrument of the astronomer, but it needs a little help to get a close look at the universe.

Probably the easiest and fastest way for anyone to get involved in observational astronomy is to pick up a good pair of binoculars. For the beginning amateur trying to decide on the type of telescope best suited to his or her needs, a pair of binoculars offers a convenient way to become acquainted with the sky. After spending some time learning the constellations with the naked eye, good binoculars can show you the full extent of what lies just beyond the reach of the human eye.

The stereo view the binoculars give you makes the stars stand out in 3-D. Because you do not strain your eyes, you are a more relaxed observer. Besides, using binoculars is a natural way to bridge the gap between naked-eye observing and the use of a telescope of any size. Using binoculars allows the beginner to develop the skills necessary to get the most out of a telescope.

It isn't a bad idea for the more advanced amateur to pick up a pair of binoculars, too. After crouching at a telescope for any length of time you can forget the grand, sweeping vistas offered by binoculars. Using binoculars in conjunction with the telescope will also allow you to detect faint objects, then determine the best way to position the main telescope for a better look.

A

B

C

Fig. 3-1. Binoculars come in all shapes and sizes. From the wide-angle 7×35s (a) through the standard 10×40s (b) up to the giant 11×80s (c) any pair of binoculars can give you an entree to the stars. (Photos a and b courtesy Bushnell/Bauch & Lomb. Photo C courtesy Meade Instruments Corporation.)

BINOCULAR FACTS

Binoculars come in all shapes and sizes (Fig. 3-1). The light path in a pair of binoculars does not run straight through the instrument as it does in a refractor telescope. If it did, the resulting image would be upside down and very hard to interpret. Binoculars make use of prisms that correct the orientation of the image and make the light path from the objective to the eyepiece short and easily managed. The *porro prism* (Fig. 3-2) type of binoculars uses two side-by-side prisms to bend the light and correct the image. The *roof prism* (Fig. 3-3) type uses a more compact set of prisms to direct the light. Roof-prism binoculars are lighter and more compact than those using the more conventional porro prism.

Binocular size is defined by two numbers separated by a ''×'' symbol. Examples of different size binoculars are: 7×35, 7×50, 10×50 and 20×80. The first number is the magnification of the instrument; the second is the size of the objective. The magnification is the number of times larger the image will appear than if viewed by the naked eye. A 7×35 instrument will yield an image 7 times normal size, while a 20×80 binocular will give you an image 20 times larger than that seen by the eye alone. One important thing to remember is that the higher the magnification of any optical instrument, the more difficult it is to keep them and the image steady.

Binoculars up to a magnification of 10 are okay to hold in your hand, but if you want more powerful binoculars you will have to use them with some type of mount. Many of

Fig. 3-2. Light path through a porro-prism binocular. (Courtesy Bushnell/Bausch & Lomb.)

the more powerful binoculars, such as 11×80 and 20×80, come with either a separate bracket or built-in socket for mounting them on a photo tripod (Fig. 3-4).

Binoculars enable the amateur astronomer to increase the light gathering power of the eye. The most important factor in determining the amount of light taken in by an optical system is the size of the area that receives the light. In the eye this receiving area is the pupil, in the binoculars it is the objective lenses. If you want to determine the comparative light gathering power of two lenses you have to compare the size of the two areas available to receive the light. If the size of a lens is doubled, its area increases four times. In other words, the bigger the lens the more light it can gather.

If a lens is 25mm in diameter it has an area available to receive light, or a surface

Fig. 3-3. Light path through a roof-prism binocular. (Courtesy Bushnell/Bausch & Lomb.)

Fig. 3-4. Many larger pairs of binoculars come with a tripod adapter to help hold them steady. (Courtesy Bushnell/Bausch & Lomb.)

area, that is about 1,550 square mm. If you want to compare its light gathering power to a lens that is 50mm in diameter, you have to compare the surface areas of the two lenses. The 50mm lens may be twice as big as the 25mm lens, but its surface area is 6,200 square mm or four times the surface area of the 25mm lens. So if we compare the light gathering ability of a pair of 10 × 50 binoculars to that of an average observer with a 7mm pupil we find that the binocular lens has a surface area over 50 times larger than the pupil.

This means that you can see much fainter stars when you use binoculars. I use two pairs of binoculars to help with my observing: a 7 × 35 and a 10 × 50. With the 7 × 35 I can easily see stars to the seventh magnitude and with the 10 × 50 instrument I can reach the ninth magnitude.

Binoculars have varying fields of view. It is important to know how large an area of sky you can take in with any particular instrument. All binoculars have an expression of field size printed on them. In most cases this will be a statement like "357 feet at 1,000 yards." The field of view is printed on both the instruction book and the instrument casing. In order to find out the size of the field of view in degrees, divide the number of feet at 1,000 yards by 52 because at 1,000 yards 52 feet equal one degree. So if your binoculars have a field of 357 feet at 1,000 yards, they have a 7 degree field of view. Increasing the amount of magnification will not increase the size of the field of view. In fact it will make it smaller by restricting the viewing area of the instrument.

When you are observing faint objects in the night sky it is important to have as bright an image as possible. The objective lenses of the binoculars contribute most of the power required to create a bright image, but you still need a little help from the binocular eyepieces.

The *exit pupil* created by the eyepiece looks like a little cone of light extending away from the eyepiece. The size of the exit pupil is important to the brightness of the image that you see. Most people have a normal, dilated pupil size of 7mm. For the brightest image the exit pupil should be as large or larger than the pupil size, so that the light will fill the entire pupil. The exit pupil can be easily figured for any type of binoculars by dividing the objective size by the magnification. A pair of 7×35s have an exit pupil of 5, as do a pair of 10×50s.

SELECTING AND TESTING BINOCULARS

Selecting a pair of binoculars is not as easy as one might think. Binoculars can serve a multitude of functions. In addition to being used to observe the stars, they can go to sporting events or on nature field trips. If you intend to use them for such things it is best to get a general purpose pair like 7×35s, 7×50s, or 10×50s. Many manufacturers are now making large objective instruments like 11×80s, 20×80s, and even 25×120s. These are special types of instruments and are really only suited for astronomical use. Regardless of the size you decide on, remember to get a pair with a large field of view and an adequate exit pupil.

There are a number of steps you can take when purchasing a pair of binoculars to make sure you are getting a quality instrument. First, check the alignment of the binoculars. The barrels should point in the same direction. If they don't you will see a double image for a brief instant while your eyes correct the problem. If you suspect poor alignment focus the binoculars on an object and then draw them slowly away from your eyes to a distance of a few inches. The image should remain single. Next, with the binoculars still focused, slowly blink your eyes a few times. If the pair is out of alignment, it will take your eyes a couple of seconds to realize the images are not properly aligned before they correct for the misalignment. Using a pair of binoculars with this defect will result in eye-strain and a headache.

The barrels should flex freely but not loosely as you move them together. Give the barrels a slight twist. There should be no play in the hinge between them. The focusing mechanism should work smoothly. The eyepiece frame should remain stable and not wobble when you are racking the eyepieces in and out.

Now stand with your back to a bright light and let it fall through the objective lens onto the inside of the barrel. The interior surfaces should be clean. The reflections you see on all the optical surfaces should be purple or amber. This indicates that the instrument has anti-reflective coatings that will increase light transmission and contrast.

Turn the binoculars around and hold them up to a bright surface. You will see a small disk of light "floating" in each eyepiece. These are the exit pupils. It is here that you can best tell if the manufacturer is skimping on quality. The exit pupils should be complete, round disks. If they are, the prisms used in the binoculars are the right size for the instrument. Many manufacturers are trying to get by with smaller and cheaper prisms.

These inferior prisms give the exit pupil a squared-off look because they cut off some of the light. The new BAK-4 prisms give the best exit pupil definition.

Images seen through the instrument should be bright and sharp. Gray or filmy images indicate that there is a problem with the contrast in the lens set. The field of view should be wide and bright. Check the quality of the field edge by moving the instrument over a straight line, like a roof edge or a telephone line, at a right angle. Notice if the line tends to distort as it moves into and out of the field of view. Take a look at a line dividing a bright area from a dark area. Because no lens surface is totally free of chromatic aberration you will see some red and blue along the line. The less you see of these colors the better the instrument.

Finally, look at a brick wall. The image of the bricks should be equally focused across the field of view. In an inferior pair of binoculars the center is sharply focused but the edges may appear bowed and curved. This indicates that the binoculars' field of view is not flat. Using them over an extended period of time will lead to fatigue.

It is important to get a quality pair of binoculars, but you cannot judge quality solely on price. Prices may vary greatly. A pair of 7×35 binoculars may range in price from \$24 to almost \$200. The manufacturers of some expensive binoculars take shortcuts in production that would surprise you, while some seemingly cheap instruments are sturdily built with good optics. When you are deciding on a pair of binoculars look for ones that have the best optics and largest aperture and still fit your budget.

Regardless of price or size you must remember that it is how the instrument is used that is important. Great discoveries are not made by instruments, but by the observers using those instruments.

OBSERVING WITH BINOCULARS

The first thing you need to remember about using binoculars for astronomical observation is comfort. Standing with your neck craned, holding binoculars over a long period of time can become very tiring. If you have to stand, it is best to mount the binoculars on a photo tripod or some other kind of stand.

You can even take binocular observation a step further and build an observing stand that allows you to look down with no strain. Such a stand is pictured in Fig. 3-5. It is constructed using a simple first-surface, flat mirror that reflects the light from the stars up into the binoculars' objectives. Using a stand like this allows you to relax and observe from a seated position (Fig. 3-6). The best way to observe is from a reclining position. A reclining yard chair with arms gives you a relatively stable platform from which to observe.

A good object to look at through your binoculars is the Moon. With good 7×50 or 10×50 binoculars, you can follow the line of sunrise as it crosses the lunar surface. The *terminator* throws the Moon's features into sharp relief allowing you to study most of the larger craters.

We've talked about the advantages that using binoculars give an observer, but what can you see with them? Good binoculars will show you all the planets of the Solar System with the exception of faint Pluto. If you have optically excellent binoculars you will be able to make out the phases of Venus as it moves around the Sun. The four bright satellites of Jupiter are easily seen. If you have a steady hand or a stable mount you may be able to see the rings of Saturn. Mercury, Mars, Uranus, and Neptune, unfortunately, will only show up as points of light because these planets are either very small or very far away. The best thing for you to do is begin by studying the constellations using the greater light grasp of your binoculars.

If you live in the city, or are bothered by severe light pollution, the additional universe that a good pair of binoculars puts at your disposal will astound you. After taking some time to reacquaint yourself with the new views of the constellations that your pair of binoculars is able to give you, it is time to meet a few of the denizens that inhabit the great bowl that lies above our heads.

In early July Lyra (Fig. 3-7) is almost directly overhead so it is a hard constellation to miss. The brightest star of the group is Vega, the fifth brightest star in the sky. Trailing south from Vega is a grouping of two third- and two fourth-magnitude stars in the form of

Fig. 3-5. Amateur astronomers are an adaptive lot. A sky scanner built to specifications found in a monthly astronomy magazine allows the observer to look down instead of bending over backwards.

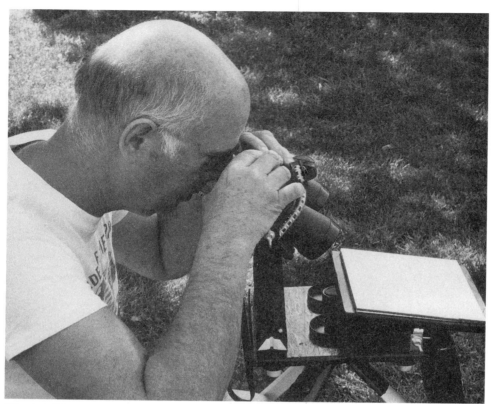

Fig. 3-6. Using a sky scanner the observer can look at the sky from a relaxed, seated position.

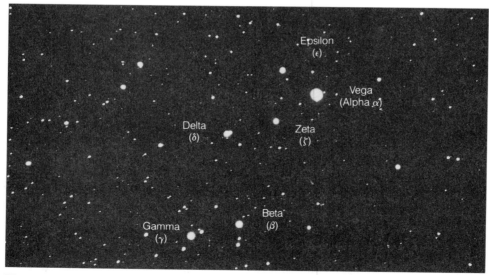

Fig. 3-7. The constellation Lyra. (Courtesy Walter Piorkowski, ©1988.)

a parallelogram. These make up the main body of the constellation. Turn your binoculars on the star at the northeastern corner of this figure and you will find yourself looking at Delta (δ) Lyrae. You will see two stars, one red and the other blue. Delta is what is known as a *double star*—two stars held together by gravity and revolving around a common point in space.

Directly opposite Delta Lyrae is Beta (β) Lyrae. If you watch Beta over a period of about 12 days you will notice that it is not constant in brightness. Beta is a *variable star*. It varies in brightness from magnitude 3.4 to magnitude 4.1. Compare Beta to the star at

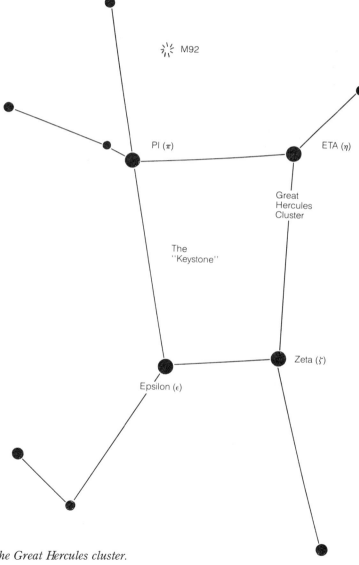

Fig. 3-8. Location of the Great Hercules cluster.

the end of the parallelogram, Gamma (γ). Gamma shines at magnitude 3.25. When Beta is at its *maxima*, or brightest, it is almost as bright as Gamma. At its dimmest, or *minima*, it shines at only 65 percent of Gamma'a brightness. Beta's variability is caused by a companion star that moves around it. Every 12 days Beta is eclipsed by its companion causing the light we see to dim.

Moving directly west from Lyra, you will come across another faint constellation, Hercules (Fig. 3-8). Four stars, three of third- and one of fourth-magnitude, make up the identifying Keystone asterism, or star grouping, of the constellation. Our next target is located between two of those magnitude 3 stars, Eta (η) and Zeta (ζ). Between them you will find probably the best example of a *globular cluster*—the Great Hercules cluster. In my 10×50s, the Great cluster appears as a soft glow surrounding a bright center (Fig. 3-9). The cluster is a grouping of hundreds of thousands of stars packed together so tightly that its center cannot be resolved into individual stars even by the largest telescopes.

Under my Chicago skies, Vulpecula is a barely visible constellation just south of Cygnus (Fig. 3-10). It passes overhead in early September. Turning your binoculars on the area shows you a wealth of detail that would be invisible to the eye, even under the best of skies. The Milky Way is rich in the area. What was formerly an area marked by two fourth-magnitude stars becomes choked with the brightness and diversity of our galaxy.

Amid this multitude of stars, the very fine *planetary nebula* known as the Dumbbell nebula becomes visible (Fig. 3-11). Nebulae like this are the result of a star blowing its outer surface into space. The gasses and matter expelled form a shell that slowly moves out from the star into space. In most cases the star that suffered the explosion is invisible to all but the largest telescopes. All we get to see is the beauty of the aftermath. With

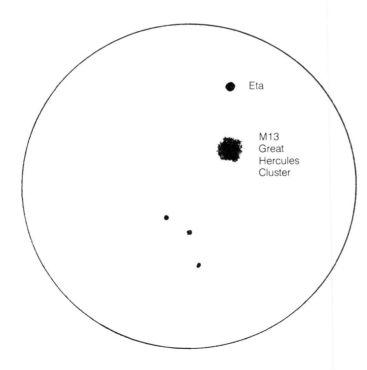

Fig. 3-9. As seen through a 10×50 binocular the Great cluster (M13) appears as a bright "smudge" of light against the sky.

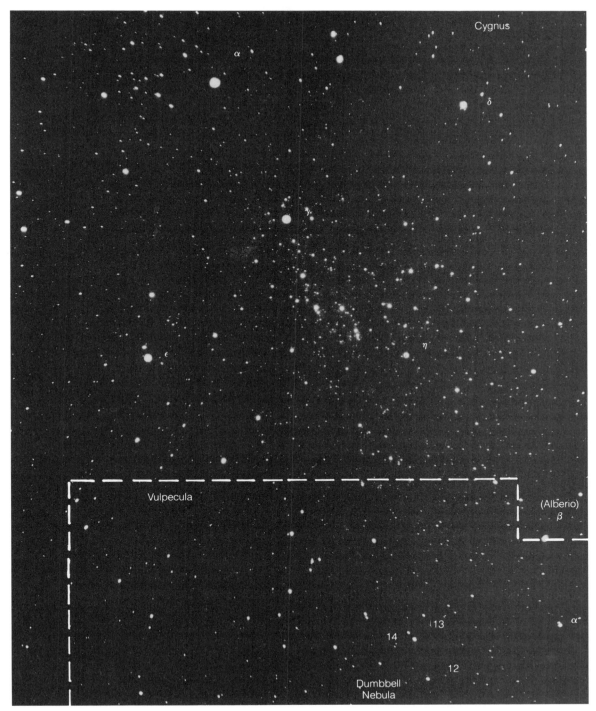

Fig. 3-10. The Cygnus/Vulpecula region of the sky. (Courtesy Walter Piorkowski, ©1988.)

the added contrast of the binoculars producing a darker sky, the nebula floats ghostlike in the field.

Another of our denizens of the deep sky is found in the autumn constellation of Andromeda (Fig. 3-12). The main asterism of the constellation is made up of two arcs of stars. The eastern arc has two second- and one third-magnitude stars and the smaller, fainter, more westerly arc made up of three fourth-magnitude stars. Both arcs are linked by the magnitude 2 star Alpha (α) Andromedae. To locate our next target find the middle star of the eastern arc, Beta (β), and move west past the second arc. In your field you

Fig. 3-11. The Dumbbell nebula. (Courtesy U.S. Naval Observatory.)

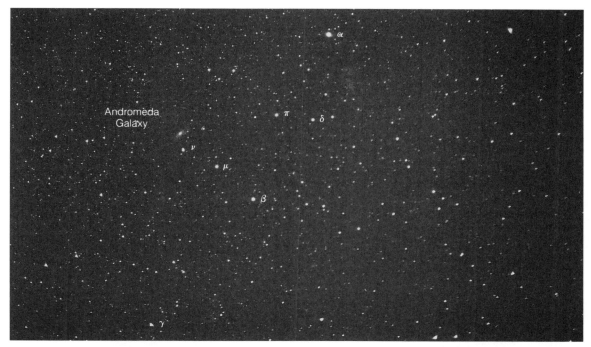

Fig. 3-12. The constellation Andromeda. The Andromeda Galaxy is visible as just above Nu (ν). (Courtesy Walter Piorkowski, ©1988.)

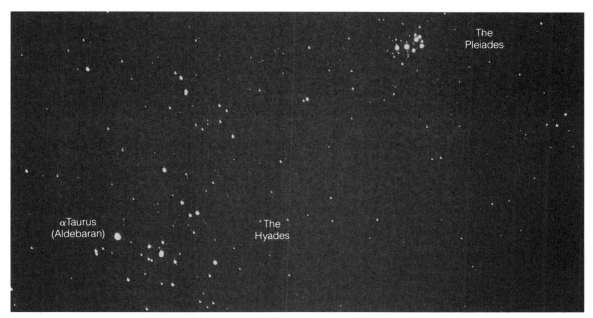

Fig. 3-13. The constellation Taurus is easy to pick out because of the Pleiades and the Hyades. These star clusters are even visible to the naked eye from a city like Chicago. (Courtesy Walter Piorkowski, ©1988.)

will see a faint, greenish glow. That is the Andromeda Galaxy. Similar to our own galaxy, the Milky Way, Andromeda is 2 million light years away from us. While this may seem far, it is the closest *spiral galaxy* to us.

A few hours after Andromeda passes overhead the constellation Taurus becomes visible. With the red first-magnitude star Aldebaran gleaming and the Pleiades splendidly displayed just above the V of the Hyades, Taurus is easy to locate. The Hyades and Pleiades (Fig. 3-13) are two examples of *open clusters*. Because of the size of these magnificent star clusters the only way to take in all of their beauty is with a pair of binoculars. Both of these objects are groups of stars moving together through space.

Fig. 3-14. The easiest constellation to see is Orion. (Courtesy Walter Piorkowski, ©1988.)

Southeast of Taurus is probably the most recognizable constellation in all the sky. Orion (Fig. 3-14), the hunter and eternal adversary of Taurus, holds many deep-sky treasures. Hanging from Orion's belt is his sword and near the tip of this is a glow that is easily seen with the naked eye—the Orion nebula. If you turn your binoculars on this object you will see what appears to be clouds of gas embedded with stars and that is just what you are seeing. The Orion nebula is a *reflection nebula*. The colors and texture you see are lit by the stars that are glowing inside the nebula. Unlike a planetary nebula, which is created by a dying star, reflection nebulae are the birthplaces of stars.

These are the basic types of deep-sky objects that a pair of binoculars will reveal to you. Now that you have a taste of the vastness and beauty of the visible universe it is time to take the next step in our journey. Let's look at some of the tools used by astronomers.

Tools of the Astronomer

There is no greater relaxation than to embark on a celestial sight-seeing trip with a...telescope.

William T. Olcott

The time will come when you will want to look deeper into the universe. For that you will need a telescope. It is the telescope that really allows the eye to see the wonders of the universe. But before you go out and plunk down your hard earned money you will need a few facts. It is hard to take advantage of an informed consumer.

A glance through any of the popular astronomy magazines yields a seemingly unending number of telescope shapes, sizes, and varieties. But when all the fancy trappings are stripped away there are only two basic types of telescopes. They operate using the same principles (Fig. 4-1). Both collect light and produce an image where the collected light comes to a focus. The part of the telescope that collects light is called the *objective*. In the refractor telescope the objective is a lens; in the reflector telescope it is a mirror.

The distance it takes for the objective to form an image is called the *focal length* of the telescope. Directly related to the focal length is the *focal ratio* (f number) of the telescope. The focal ratio of the instrument can be found by dividing the focal length by the objective diameter. These figures can be expressed either as inches, millimeters, or centimeters as long as the same unit of measure is used throughout the process. Most manufacturers measure focal lengths using the metric scale.

A 6-inch objective with a focal length of 30 inches has a focal ratio of f/5. If it had a focal length of 48 inches its focal ratio would be f/8. The smaller the focal ratio the shorter the focal length and the faster the system. The larger the focal ratio the longer the focal length and the slower the system. A 6-inch f/5 is faster than a 6-inch f/8. A 6-inch f/10 is slower than both of them.

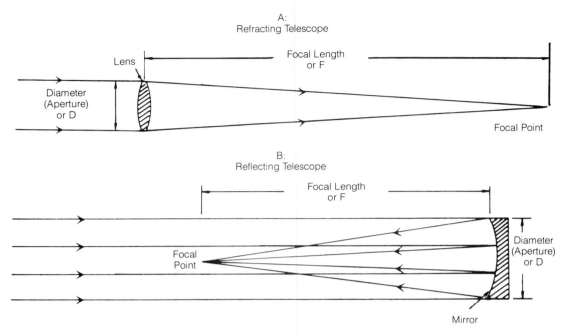

A:
Refracting Telescope

Lens

Focal Length
or F

Diameter
(Aperture)
or D

Focal Point

B:
Reflecting Telescope

Focal Length
or F

Focal
Point

Diameter
(Aperture)
or D

Mirror

Fig. 4-1. Telescopes operate by bending light. A refracting telescope (a) uses a lens while a reflecting telescope (b) uses a mirror. (Courtesy Tasco.)

This speed designation is useful when you consider what you want to observe with your telescope. The faster the system, the larger and brighter the field of view. Fast systems, like f/4 or f/5, are ideal for looking at very faint, diffuse objects like star clusters, nebulae, and galaxies. Slower systems, f/10 and up, give better contrast in the field of view and excellent views of bright objects like the Moon, planets, and double stars.

FUNCTIONS OF A TELESCOPE: LIGHT GRASP

The most important aspect of any astronomical instrument is its ability to act as a light collector. This is known as the telescope's light gathering power or *light grasp*. The light gathering power of a telescope is a function of the size, or *aperture*, of the objective. The light gathering power of the objective is compared to that of the fully dilated pupil of a healthy eye, which has an aperture of 6mm or 7mm.

As we have seen with binocular objectives, the light gathering power of the telescope objective increases as its size increases. So the light gathering power of a 4-inch objective is not twice, but four times, that of a 2-inch. Although the diameter of the objective has only doubled, the area of the objective has increased four times. The total area of the objective collects light. The larger the objective, the greater the light grasp of the instrument and the fainter the objects it can see. Table 4-1 shows you the comparative light gathering power of the human eye and various size objectives.

Aperture (mm)	(In)	Light Gathering Power (Eye = 1)
35	1.4	25
50	2.0	50
60	2.4	80
80	3.1	130
100	4.0	210
150	6.0	480
200	8.0	840
250	10.0	1300
317.5	12.5	1900
330	13.1	2100
355	14.0	2400
405	16.0	3360
444	17.5	3680
500	20.0	4900

Table 4-1. Light Gathering Power

Limiting Magnitudes

The faintest star the unaided eye can see under perfect conditions is sixth magnitude. With the increased light grasp of even a small 50mm telescope you may see stars almost 100 times fainter. But no matter how big a telescope is, it can only see stars of a certain brightness. A telescope still has a limiting magnitude.

The limiting magnitude of a telescope is the faintest star that can be seen through the telescope under ideal conditions. The limiting magnitudes of various size telescopes are listed in Table 4-2. These numbers do not, however, take into account such factors as the visual acuity or skills of the observer, and they have been exceeded by more than one observer. A quick look at Table 4-2 however, would tell an amateur astronomer with a 60mm refractor telescope that it would be futile to try to locate the planet Pluto because it is usually found around magnitude 14, well past the limit for the instrument.

Table 4-2. Limiting Magnitudes

Aperture (mm)	(In)	Limiting Magnitude
35	1.4	9.5
50	2.0	10.5
60	2.4	10.9
80	3.1	11.5
100	4.0	12.0
150	6.0	12.9
200	8.0	13.5
250	10.0	14.0
317.05	12.5	14.5
330	13.1	14.6
355	14.0	14.7
405	16.0	14.8
444	17.5	15.0
500	20.0	15.5

Image Brightness

The objects that amateur astronomers observe are of two types. Stars are *point source objects*. No matter how big a telescope you use they will always be seen as points of light. The Moon and most planets, on the other hand, show a definite disk when viewed through even a small telescope. Because of this they are called *extended objects*. Star clusters, nebulae, and galaxies are collectively known as *deep-sky objects*. They comprise another class of extended objects.

The fact that an 80mm refractor is capable of reaching magnitude 11.5 can be a bit deceiving. What it really means is that you may be able to see point sources, or stars, to that magnitude under ideal conditions. It does not always mean you can see deep-sky objects of that magnitude. The brightness of the image of a deep-sky object is not concentrated in one central point, but extends over the entire visible area of the object. While it may have a high magnitude rating, its overall surface brightness can be low, making it hard to see.

An object like the spiral galaxy M33 is a good example of this phenomenon. Its visual magnitude is listed as 6.7, and if you are far enough away from the annoying lights of the city you can just make it out against a dark sky. It looks like a ghostly mist covering an area about the size of the full Moon. But the magnitude that is listed for the galaxy is what is called an *integrated magnitude*; that is, it is what the magnitude of the galaxy would be if it were seen as a point source. Because M33 covers such a large area of the sky its light is spread over a large area. If you look for it through your telescope it may be impossible to find.

FUNCTIONS OF A TELESCOPE: RESOLUTION

Diffraction is a phenomenon of light that allows us to see detail in the universe. Diffraction causes light to bend ever so slightly when it encounters an obstacle, such as when it enters a telescope tube. This bending causes the resultant image of the star to distort and appear as a miniscule, blurred dot through the telescope. This image of the star is called a spurious disk or *airy disk* (Fig. 4-2).

Named for the seventh astronomer royal, Sir George Airy, the airy disk looks like a tiny bullseye of light formed by a bright central disk surrounded by alternating dark and bright rings. The disk's greatest brightness is found in the center while it drops off to almost nothing at the first dark ring, also called the first *interspace*. The brightness then increases again, although to only a fraction of the central brightness, for each discernible ring. With a 6-inch telescope and good seeing, a first-magnitude star will show as many as five or six of these diffraction rings.

The Resolving Power of a Telescope

When two equal point sources of light are close together, as with a double star, the airy disks overlap and the images are combined into a single airy disk (Fig. 4-3). When you are able to discern this disk into its individual components through the telescope it is said to be resolved. The threshold or resolution of an instrument is found by the formula: R = 5.35/D where *D* is the aperture of the telescope in inches. The resolution figure (R) is known as the *rayleigh limit* of the telescope. It occurs when the two images are separated by the first diffraction interspace.

William Dawes, an English amateur, noticed that this theoretical limit could be exceeded by about 5% under good conditions and redefined the resolution limit of a telescope to: $R' = 4.56/D$ where D is the aperture of the telescope in inches. Known as the *dawes limit*, it differs from the rayleigh limit in that it is the first point a which a double star is elongated enough to suspect the presence of two stars.

Like light grasp resolution is dependent on the aperture of the instrument. The only way to increase the resolving power is to increase the size of the aperture. No amount of magnification will enable you to resolve a double star that is beyond your telescope's limit. Under actual observational conditions these limits are rarely reached because a number of factors, including atmospheric stability and the visual sensitivity of the observer, can combine to keep a telescope from reaching its resolution limit.

The resolving power of a telescope is applied most often to double stars, but that is not the only area where it is important. The resolving power of a telescope dictates such things as the sharpness of detail visible on a planetary disk and the smallest crater visible on the Moon. Table 4-3 gives the rayleigh and dawes limits for various size telescopes.

FUNCTIONS OF A TELESCOPE: MAGNIFICATION

The image formed at the primary focus of a telescope is slightly magnified. The eyepiece of the telescope acts on the image and increases it to a size large enough to be examined in detail. This process is commonly known as *magnification* and is the least important

Fig. 4-2. The airy disk.

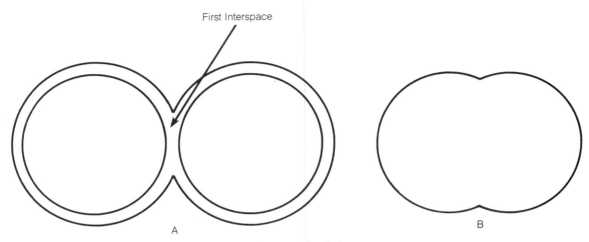

First Interspace

A

B

Fig. 4-3. A double star is said to be resolved at its traditional rayleigh limit (a), when the two airy disks are separated by the first interspace. The dawes limit (b) only requires them to be elongated enough to tell there are two stars.

function of a telescope. Unlike the other functions of the telescope, magnification does not depend on the size of the aperture. Rather the focal length of the objective combined with the focal length of the eyepiece determines the instrument's magnification power.

To find the magnification of an eyepiece divide the focal length of the telescope by the focal length of the eyepiece. The result is the magnification power of the eyepiece. A 3-inch refractor telescope with a focal length of 910mm and a 10mm eyepiece has a magnifying power of 91 times or $91 \times$.

In order to increase the magnification you have to change the focal length of either the objective or the eyepiece. Changing the focal length of the objective would mean replacing the objective because each telescope comes with an unalterable focal length. It is easier to replace the eyepiece with one of a different focal length. This has its prob-

Table 4-3. Resolution Limits

Aperture (mm)	(in)	Resolution Arcseconds Rayleigh Limit	Dawes Limit
35	1.4	3.9	3.2
50	2.0	2.8	2.3
60	2.4	2.3	1.9
80	3.1	1.8	1.5
100	4.0	1.4	1.1
150	6.0	0.9	0.76
200	8.0	0.68	0.57
250	10.0	0.54	0.45
317.05	12.5	0.43	0.36
330	13.1	0.41	0.34
355	14.0	0.38	0.32
405	16.0	0.34	0.28
444	17.5	0.31	0.26
500	20.0	0.27	0.22

lems, however. As the magnification increases the focal length of the eyepiece becomes smaller and the eye must get closer to the eyepiece, until it is touching the lens. This is both uncomfortable and a waste of light.

It is much more important to consider the size of the exit pupil of the eyepiece. The exit pupil, also called the *ramsden disk*, is that small circle of light that seems to float above the eyepiece. It is the image of the telescope's objective illuminated by the object at which you are looking. It should be just as large as the pupil size of your eye. This allows it to transmit all the light gathered by the telescope to your eye so you can see the brightest image. On the other hand, if you are observing a very bright object such as the Moon, your pupil will contract and reduce the amount of light falling on the retina because the extra light isn't needed to form a sharp image. So a smaller exit pupil will suffice to fill the smaller size of your eye.

The exit pupil imposes a range of magnification on a telescope and it is best if the user stays within that range. If the exit pupil of the eyepiece is so large that the eye's pupil cannot take in all the light the telescope is working at a disadvantage. The full light gathering power of the objective is not being used. In effect, you are decreasing the size of the telescope.

It is generally accepted that the lower limit of magnification is three or four times the diameter of the objective in inches (3D or 4D). A limit lower than this would rob the telescope of its ability to resolve an image sufficiently. Using this formula, a 3-inch refractor telescope would have a lowest useful magnification limit of 9 to 12 power. If the focal ratio of the instrument was f/11 the desired exit pupil would be between 6 and 7mm.

The highest allowable magnification is generally accepted to be 25 times the aperture of the objective (25D). In our example, this would give a magnification of 75 for a 3-inch with an exit pupil of 1mm. Using too much magnification is self defeating. After a certain point, the only thing magnification will do is reduce the sharpness of the image.

All this talk of exit pupils and magnification ranges should not scare you. Like anything else there are exceptions to the rule such as specific observation situations. Certain objects take magnification better than others. Double stars, provided they are above the instrument's resolution threshold, sometimes need the maximum magnification you can get from your instrument in order to separate them. Observing the planets, with their wealth of detail, also calls for higher magnification to make the most of the telescope's resolution ability. In some instances magnifications as high as 50 or 60 times the objective can be used, but those circumstances depend on steady eyesight and the type of object you are observing. Double stars and planets can take a lot of magnification.

OPTICAL IMPERFECTIONS

No optical instrument is perfect. This is as true for the telescope as for the eye. All telescopes are subject to five types of imperfections called *aberrations*. No matter how much an observer pays for an instrument, or how long he works on it if he makes it himself, it will never be entirely free of imperfections.

Refractor telescopes work on the principle of bending light to a point of focus. Because of this they tend to act like a prism. The white light that enters the optical system separates into its component wavelengths by the time it arrives at the focus. In a system with a simple lens one image is formed by the red light inside the point of focus while another image is formed by the blue light just outside the focus. This is called *chromatic* aberration (Fig. 4-4) and is the bane of the refractor. Looking at anything in a system like this would tend to cause headaches because everything is surrounded by halos of color. Reflectors don't have this problem because the mirror reflects all the light, regardless of wavelength, equally.

All optical systems are subject to other forms of aberration. *Spherical* aberration (Fig. 4-5) is caused when the light from a lens or mirror does not reach the same focus. An image produced by a mirror or lens with spherical aberration will appear as a series of overlapping images. *Coma* (Fig. 4-6) is an aberration that causes the images of a star to look like a tiny fan. Finally, *astigmatism* (Fig. 4-7) occurs when the lens or mirror is unable to bring the image to a single focus. Instead the image is formed in front of and behind the point of focus and appears blurred and indistinct.

TELESCOPE TYPES: THE REFRACTOR

When you say "telescope," the image that pops into most people's minds is that of the refractor (Fig. 4-8). The smooth, long lines of a massive refractor telescope sitting on its pier and mount in an observatory has long been the symbol of astronomy to most people.

The refractor has changed quite a bit since Galileo's time, but the operating principle has remained the same. Galileo used a simple lens system. The first lens gathered the light and focused it; then a second lens took up the duties of focus and magnification. His telescope was small, never achieving a magnification over 30 power. Today, a fairly straight line descendent of Galileo's telescope survives in the common field glass that uses a simple lens and eye lens.

A telescope that uses a single, simple lens has one major flaw: chromatic aberration. Chromatic aberration gives the image a halo effect—bright objects surrounded by rings

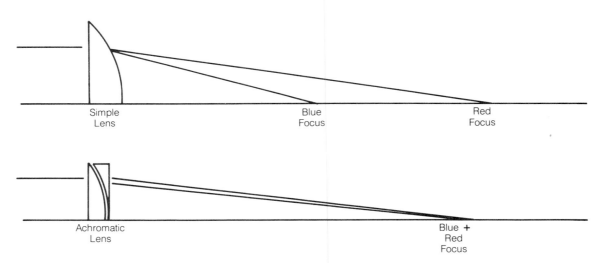

Fig. 4-4. Chromatic aberration brings light of different wavelengths to different points of focus. A doublet, or achromatic lens, corrects this.

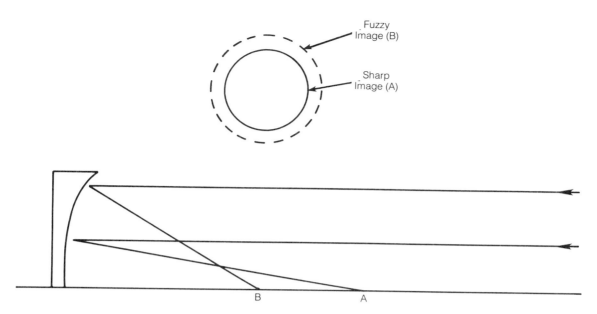

Fig. 4-5. Spherical aberration causes part of the image to come to a sharp focus while the rest of the image is soft and fuzzy.

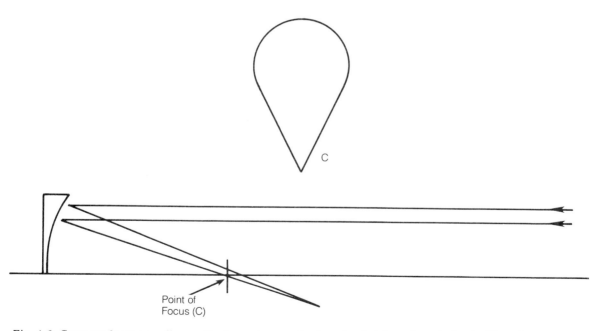

Fig. 4-6. Coma produces a small, conelike image because the focus is offset from the optical axis. The point of the cone always points toward the optical axis.

of color. This is caused when the differing wavelengths of light cannot be brought to a single point of focus.

In the early days of astronomy this problem was overcome through the use of telescopes with extremely long focal lengths. Chromatic aberration can be minimized in a simple lens by using a focal length that moves the focal point farther away from the lens. Charles Messier, the astronomer who cataloged a list of objects that bears his name, used telescopes with focal lengths of 60 feet. The astronomer Hevelius built a telescope 150 feet long, so long that he could barely use it!

In 1757 the English physicist John Dolland discovered the basic principle of the *achromatic lens*. When two lenses made of different types of glass were placed together each element canceled the chromatic aberration of the other. Working in the early 1800s, the German physicist and astronomer Joseph von Fraunhofer discovered that chromatic aberration could be further reduced if the rear surface of the front element was more strongly curved than the front surface. He also perfected the spacing of the two elements and the *achromat*, or modern telescope lens, was born. Lenses produced today could have been designed by Dolland and Fraunhofer.

State-of-the-art optics today call for the use of another lens in the design of refractor objectives. This creates a lens with three elements called *apochromat*. With the apochromatic lens, chromatic aberration is reduced to the point where it is no longer a factor in the telescope's operation.

The refractor telescope available to today's amateur is a vast improvement over the instruments used by some of the pioneering observers of the science (Fig. 4-9). They require little if any maintenance, and their sealed tubes keep out dust and air currents

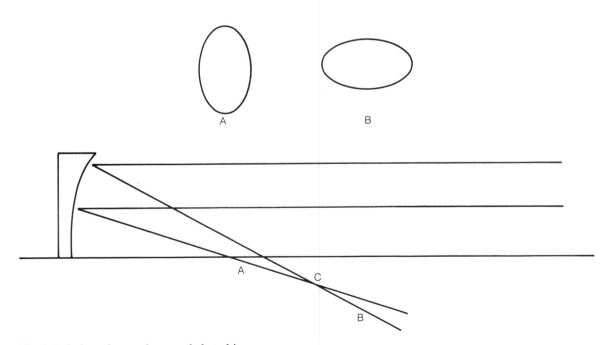

Fig. 4-7. Astigmatism produces oval-shaped images.

that might affect the image. The alignment or collimation of the optical system is simple to adjust and relatively permanent.

Refractors have always been considered best for observing the Moon, the planets, and double stars. This is because the longer focal lengths found in most refractors make for a very dark field of view and allow for the use of high magnification. These factors all contribute to the crisp, sharp images of bright objects that can be seen through the refractor.

On the other hand, refractors do not provide good views of deep-sky objects. Deep-sky objects and objects with a low surface brightness require the use of instruments with faster focal ratios and brighter fields. A number of companies make refractors with small, faster focal ratios. A telescope with a 5- or 6-inch aperture and a short, fast focal ratio is considered the prime visual instrument for comet hunting. Leslie Peltier, the famous amateur astronomer, discovered 10 comets using a 6-inch f/8 refractor telescope.

One of the major drawbacks of the refractor telescope is its cost. A good instrument with a well-corrected, 80mm (3.1-inch) objective and good, sturdy mount will cost over $175 per inch. There are a number of companies who produce what has become a standard instrument in the refractor class. This is the 60mm, an instrument that introduced

Fig. 4-8. A refractor like this is the visual image that most people attach to the word telescope. (Courtesy David Bachtell, Southeastern Iowa Astronomy Club, ©1988.)

many amateurs to the wonders of the heavens. Unfortunately, some of these manufacturers are less than strict when it comes to quality control. These instruments can be found at toy stores, department stores, and discount houses across the country for between $120 and $150. The amateur who is looking at an instrument like this would be wise to keep a wary eye on the quality of the instrument, especially its mount. A telescope will be no good to you if the mount is wobbly.

TELESCOPE TYPES: THE REFLECTOR

About 56 years after Galileo introduced the world to the wonders of the universe with his telescope another great scientific mind was considering how to improve on that work. Sir Isaac Newton knew about the chromatic aberration caused by the unequal focus of the refractor lens. He also knew that one of the properties of a mirror was the equal focus of all the light striking it, regardless of the light's wavelength. He devised an instrument that took advantage of that fact and created the instrument we know today as the newtonian reflector (Fig. 4-10).

The reflector telescope uses a mirror to bring light to a point of focus. The surface

Fig. 4-9. A commercially made 80mm refractor. (Courtesy Meade Instruments Corporation.)

of the mirror has to be a specific shape—a paraboloid—in order for this to work properly. Amateur astronomers have been producing homemade mirrors for their telescopes for years. It is a fun and easy project for anyone interested in combining craftsmanship and science.

Because it is a simple instrument and easy to build, the reflector telescope is the favorite of the amateur astronomer. Newtonians are readily available from a number of manufacturers. They come in a much wider range of aperture sizes than the refractor and, owing to their ease of manufacture, are much more reasonable in cost. A newtonian with a 4.5-inch (114mm) mirror and a solid equatorial mount can be purchased for about $125 per inch.

Reflector telescopes are free of the chromatic aberration that plagues the refractor telescope but they have their own set of optical problems. The parabolic surface of the objective, called the primary mirror, must be made to a very fine set of tolerances. The mirror used in a reflector is subject to the same optical imperfections as a lens with the exception of chromatic aberration. The major aberrations found in reflectors are spherical and coma. These are caused by poor optical workmanship during the figuring of the mirror. The quality of the mirrors used by commercial telescope makers are good, but the constraints of the modern assembly line do not facilitate the care and time that most amateur telescope makers lavish on their instruments.

There are two major types of reflector telescopes: the newtonian and the cassegrain. In the newtonian (Fig. 4-11) light from the primary mirror is reflected off a second

Fig. 4-10. Walter Piorkowski's 16-inch reflector is an example of this instrument's wide range of sizes. (Courtesy Walter Piorkowski, ©1988.)

mirror, called the *diagonal*. Light travels from the diagonal into the eyepiece and then into the eye. Each time the light bounces off one of these surfaces a little of it is lost. Incoming light must also travel around the diagonal mirror and its supports, which causes more light loss. The amount of light that finally reaches the eye can be anywhere from 65 to 85 percent of the light that first entered the telescope. The focal length of a newtonian reflector is normally between f/5 and f/8. At the low end, an f/5 telescope makes a good deep-sky telescope if used under a dark sky. The long and more traditional f/8 gives a darker field of view and allows for a much sharper image.

Both the newtonian and cassegrain systems use a figured paraboloid for the primary mirror. The cassegrain, however, has a secondary mirror that reflects the light back toward a hole drilled in the center of the primary mirror. This is called a folded system because the light path is folded back onto itself. The secondary mirror in the cassegrain is not the flat type that one finds in the newtonian. It is a specially figured convex mirror that acts to increase from three to five times the focal ratio of the system. A cassegrain telescope with a primary mirror of f/5 becomes an instrument with a final focal ratio between f/15 and f/25 depending on its secondary mirror. This makes for a very compact telescope. If the primary is a 6-inch mirror, a f/20 focal ratio would be 120 inches long. That would be an awfully long tube if it was a newtonian, but the cassegrain system folds that into a tube only 30 inches long.

Another popular variation on the reflector telescope is the instrument known as the rich-field-telescope (RFT) (Fig. 4-12). Built with focal lengths faster than f/5, these instruments give breathtaking views of deep-sky objects like star clusters and gaseous nebulae. Because of their fast speeds these instruments require a large diagonal mirror, which causes a greater than normal light loss. The popular dobsonian design of the 1980s combines the speed of an RFT with a large mirror and smooth mounting to make for a good instrument (Fig. 4-13).

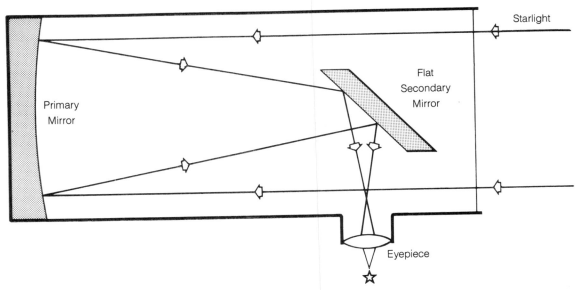

Fig. 4-11. Light path through a newtonian reflector. (Courtesy Cosmic Connections, Inc.)

Telescope Types: The Catadioptric

The *catadioptric*, or *cat,* has been called the newest wave in telescope optics. The catadioptric combines the mirror system of the reflector and a lens that acts as a corrector plate to eliminate imperfections in the mirror (Fig. 4-14). These instruments have been commercially available since the 1970s, but amateur telescope makers have been making catadioptrics since the 1940s.

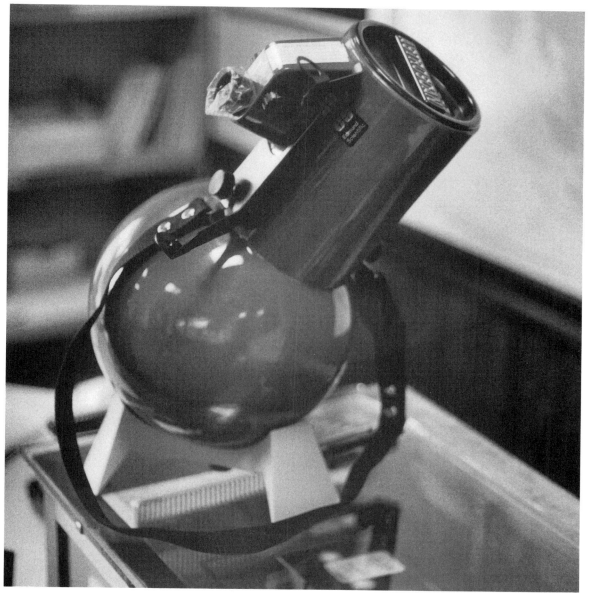

Fig. 4-12. A commercially produced 4.25-inch Rich-Field-Telescope (RFT).

Fig. 4-13. Dobsonian telescopes reflect a growing trend toward large-aperture amateur instruments.

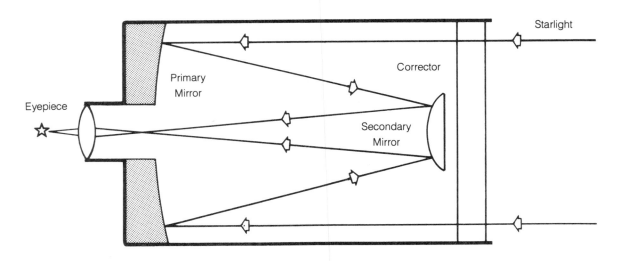

Fig. 4-14. Light path through a schmidt-cassegrain telescope (SCT). (Courtesy Cosmic Connections, Inc.)

Most of the catadioptric systems available today are known as schmidt-cassegrain (SCT) systems (Fig. 4-15). Built-in sizes ranging from 4 to 22 inches, they take advantage of the small size of the cassegrain design by using a very fast primary mirror, usually f/2, and a corrector plate to eliminate the optical aberrations found in mirrors of that speed.

The greatest advantage of the SCT is its portability. An 8-inch refractor would be impossible to store in a closet. An 8-inch reflector can fit in a closet, but not much else will fit in there with it. An 8-inch SCT, however, fits easily in a case that can reside in a corner of the closet, leaving the rest of the space free. With the SCT it is no longer a chore to take the telescope out for an evening of observing outside the city under dark-

Fig. 4-15. A 10-inch schmidt-cassegrain telescope (SCT). (Courtesy Meade Instruments Corporation.)

sky conditions. The development of the SCT has allowed the amateur to take to the roadways in search of good observing conditions.

SCTs are, in many ways, excellent all-around telescopes. If properly maintained they can give very good definition when used with bright extended objects like the Moon and planets. But because they have such large secondary mirrors, the SCT suffers from

Fig. 4-16. The schmidt-newtonian combines the speed of a newtonian and the corrector plate principle of the SCT. (Courtesy Meade Instruments Corporation.)

more light loss than a newtonian reflector. The SCT is also subject to a loss of contrast between the image and the background sky. This is because the mirror is perforated to admit the light from the secondary. This light enters the mirror along with a lot of stray light that randomly enters the telescope. This loss of contrast can be remedied by using a baffle at the mirror to screen the extraneous light. But all in all, if there is a telescope that comes close to being a universal instrument, capable of being used in all phases of astronomy, the SCT comes close. An SCT telescope on a fork-type equatorial mount will cost about $200 per inch.

A number of companies also produce a 6- and 8-inch size variation of the classic newtonian reflector using a very fast mirror combined with a corrector plate in a design called the schmidt-newtonian (Fig. 4-16). Another version of the cat is the maksutov. Smaller than other cats, available in 3.5- to 7-inch apertures, maksutous use a different type of corrector plate and primary mirror that can produce images of the Moon and planets that rival those from a refractor (Fig. 4-17).

OTHER FACTORS IN SELECTING A TELESCOPE

There is more to selecting a telescope than picking a color and kicking the tires. A quality telescope is a big investment and it is important for the amateur to consider all factors

Fig. 4-17. The maksutov telescope. (Courtesy Questar Corp.)

before making a selection. In addition to concerns about aperture size, resolving power, and the strong and weak points of individual telescope types, there are a few general questions that must be answered. The three "S"s—subject, site, and sky—all combine to help narrow the choices available to the amateur astronomer.

The choice of subject is the most important consideration. Just what do you want to look at with the telescope? Objects like the Moon, planets, and double stars require high power, sharp resolution, and a dark field. A telescope suitable for studying these objects, however, will fall short when turned on deep-sky objects. On the other hand, an instrument that gathers enough light to study the fine structure of a diffuse nebula will be woefully inadequate for viewing the Moon.

Unfortunately there is no such thing as the perfect overall telescope. Regardless of your final choice, you are going to have to give up something along the way.

The site from which you will be observing will also have something to do with the type of telescope you select. An observer living in the middle of a major city, surrounded by light pollution, is going to have a different set of observing priorities than a person living 150 miles outside that city. For the observer at the bottom of a pool of light pollution the best objects to view may be those with a high surface brightness like the Moon, planets, and brighter star images.

It is also important to consider the transportability factor of the telescope. If you will have to move the instrument to the observing site, it is best to get a light, compact telescope. If you are going to observe from your backyard and can use the garage for storage, size may not mean that much to you.

Finally, if you are going to buy a telescope get to know the dealer of your local astronomy shop. There is usually at least one in each large city. He is usually an active amateur himself and can guide you along the road of equipment acquisition by sharing his experience. If you plan to build a telescope yourself, look both to your local dealer and the members of your local astronomy club for help. They are only too pleased to see a new member join the fold and can be a great help to a beginning amateur astronomer.

TAKING YOUR SCOPE FOR A TEST DRIVE

The best way to check the performance of any telescope is to use it. It would be nice if you could take the telescope for a test drive before you buy it, but this is impractical. Once you have the telescope home and assembled, take a few minutes to get acquainted with it. Get to know how the instrument will feel while it is light outside. See how far you can reach while at the eyepiece. Are the slow-motion controls readily accessible? If the telescope has a star diagonal can you use a stool or chair to get comfortable? If you can, how far toward the horizon is this useful? Practice moving the telescope through short arcs to get the feel of its ease of motion. Your next step should be to get it outside for a few simple performance tests.

Aligning the Optics

One of the biggest bugaboos confronting the beginner is the need to align the optics of the telescope and then to keep them in alignment. After you have the telescope set up and are acquainted with it, it is time to *collimate* the optical system. Don't let that big word intimidate you. All you are going to do is make sure the light entering the telescope reaches your eyes properly.

If you have a refractor telescope it is easy. Because the refractor has only one optical element, the objective lens, it is the easiest to align. Most refractors sold today come with the objective in perfect collimation and, unless you are planning to use the telescope as a baseball bat, it will probably stay in alignment indefinitely.

A reflector telescope is another matter. Here you have two elements that must line up perfectly to pass the image along to the eye. The primary mirror is mounted in a cell and can be jarred out of alignment with a solid thump. The diagonal is suspended at the end of a single stalk of thin metal or from the vanes of a spider mount. Either mounting system is susceptible to movement or vibrations that can knock it out of alignment. The actual procedure for collimating a reflector is not that difficult and most of the work can be done in full daylight.

The first step is to get a small pinhole sight and a set of small screwdrivers and wrenches. There are three ways you can get the necessary pinhole sight. You can buy one for about $30. The main drawback to this method is the $30. You can take the lens system out of an old, high-power eyepiece you no longer use. But if you are just starting out it is doubtful you have any old eyepieces laying around, so let's move on to the third option.

The easiest way to secure this precious device is to take a few pictures of your new telescope. The canister the 35mm film came in will suit our purposes nicely, and you will have a momento to show your friends. With the canister firmly in hand, remove the lid and look at its base. There is a small dimple in the center where the canister was attached to a spur when it was manufactured. Drill a small hole, centering on the dimple. The hole should not be larger than 1/16 inch. Now take the canister and turn it over. See the lip where the lid fit? Carefully cut that off with an X-Acto knife or razor blade. The resulting lipless, hole-in-the-bottom film canister will now fit into the telescope's focuser. Our newly constructed "collimation eyepiece" is now ready for use.

With the telescope pointed at the sky rack the focuser in as far as it will go and look into the pinhole sight. You will see from the outside in: the edge of the focuser, the diagonal's holder, the outline of the diagonal itself, the reflection of the primary mirror, and the reflection of the diagonal reflected in the primary (Fig. 4-18). The first thing you want to do is center the diagonal in the telescope tube. If the diagonal is mounted on a stalk, the stalk will be connected to the focusing mount by means of a set screw. Loosen this screw and sight down the tube to position the diagonal holder in the center of the tube. A spider mount will have the holder fastened to the mount vanes by means of a bolt. You can loosen the bolt and adjust the diagonal holder in the same way.

Now go back to the pinhole for our next step. You want to position the reflection of the primary mirror in the center of the diagonal (Fig. 4-19). You may have to bend the stalk slightly to do this but don't worry, it is built to take this treatment. With a spider mount, you need to adjust a number of small setscrews to move the diagonal mirror. Make your adjustments until the primary mirror, complete with the visible section of the mirror cell, is centered in the diagonal mirror.

Now its time to deal with the primary mirror. At the back of the mirror cell you will find a number of knurled bolts, usually three (Fig. 4-20). Go back to the pinhole sight and check the position of the reflection of the diagonal. What you want to do is move this

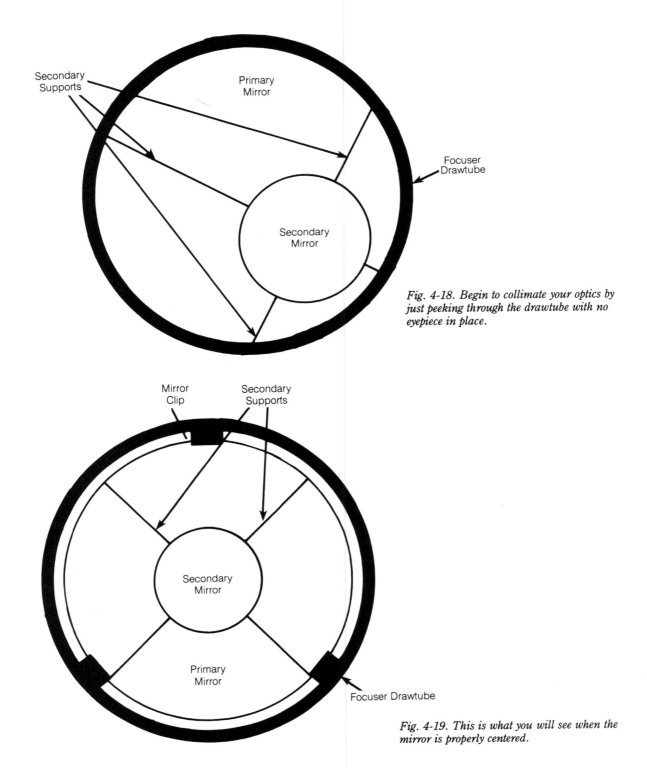

Fig. 4-18. Begin to collimate your optics by just peeking through the drawtube with no eyepiece in place.

Fig. 4-19. This is what you will see when the mirror is properly centered.

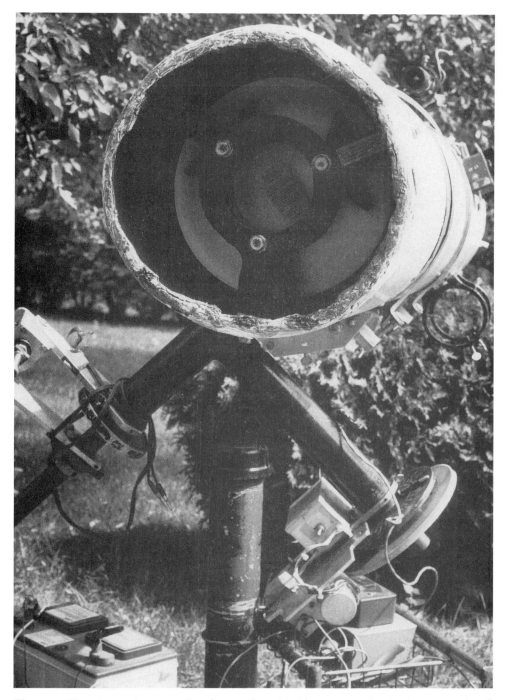

Fig. 4-20. The mirror-cell supports the mirror and can be adjusted by turning the three screws found at the back of the cell.

reflection until it is centered in the reflection of the primary mirror (Fig. 4-21). Check the image, then go to the primary and turn the appropriate screw. Repeat this until the image is centered.

The next thing to do is align the finderscope. If you aligned it before you started moving the primary you can be sure that the finder needs readjustment. You can do this during the day so you don't have to fumble with any adjustments in the dark. Select an object at least a quarter of a mile away, the farther away the better, and sight along the telescope tube until it is in the field of view of the main telescope. Put in your highest-power eyepiece and center the image in the field. Now look through the finder and see how far off the target is from the center of the crosshairs. Adjust the image until the crosshairs are centered on the target using the setscrews around the finder's mounting ring. Now select another object that is farther away and aim the telescope using the finder. If the finder is properly aligned, the target should be centered under the crosshairs and in the main field of view. If it is off center, readjust the setscrews until the target is centered. There is nothing more annoying than trying to find an object with a finder that's not properly aligned.

When the Sun has set, find a bright star and bring it into view. Now it is time to fine tune the system. Using a medium-power eyepiece throw the star slightly out of focus. You will see the star's diffraction rings. If you have properly aligned the optics, the rings will be concentric (Fig. 4-22). Most times, however, the rings will be slightly off and you will have to make a few more subtle turns on the primary cell setscrews to bring the rings into line.

SCT telescopes are easier to align. Since there is no way for you to get at the primary mirror all of your adjustments will be made with the secondary mirror. There are three small screws around the edge of the secondary mount (Fig. 4-23). These are all you will be using. If there is a large bolt in the center of the secondary mirror leave it alone! Remove the star diagonal and put a medium-power eyepiece in the focusing tube. Find a bright star and rack the focus out until a dark disk appears at the center of the out-of-focus image. This is the silhouette of the secondary mirror caused by the incoming

Fig. 4-21. A properly collimated system will show you the universe!

light from the star. The dark spot should be centered in a ring of light. If it isn't adjust the screw opposite the direction that the dark spot is off-center. For example, if the spot is left of center turn the screw to the right. After this rough alignment is made put in a high-power eyepiece and find a third-magnitude star. Center it in the telescope's field of view and slowly rack the image out of focus. The diffraction rings should be concentric. If not make any necessary adjustments to the secondary mount.

Star Testing the Telescope

It is not a bad idea to give the telescope a quick star test whenever you start an observing session. It's easy to do and it can alert you to any number of problems.

Set the telescope up in the daylight and make sure the optics are properly aligned. Check the field of view. Put in a low-power eyepiece and find an object that fills the field. The object should be in focus across the entire field. If the center of the target is crisp but the focus becomes fuzzy at the edges, the field is not flat. Switch to another eyepiece. If the problem disappears the first eyepiece has a flaw. If it persists, the optical system may still be out of alignment. The image should also be uniform. Any bulges or pinches in the image are a sign of distortion and indicate a problem.

Take a few minutes to get acquainted with the telescope in the dark. Select a second magnitude star, such as Polaris. It is a good test object because it doesn't move. Using a

Fig. 4-22. Good star images are the product of proper collimation. Star image A is what you should see with the image in focus. Images B and C are outside and inside of focus respectively.

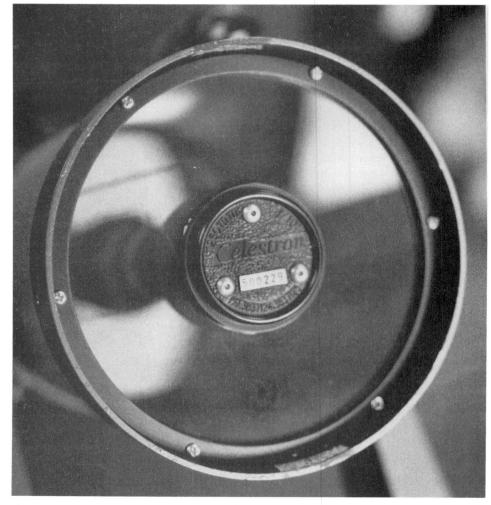

Fig. 4-23. The collimation of a SCT is accomplished by turning the screws on the secondary mounting plate.

high-power eyepiece bring the image to sharp focus and then slowly move the focus of the instrument back and forth. When it is in focus the star will form a small disk—the airy disk described in chapter 2.

As you move the focus inside or outside the focal plane the image should expand a little, but look the same at either side of the focal point. In a reflector the image formed inside the focal point should be surrounded by a series of rings—the diffraction rings. Outside the focus a dark spot should form in the center of the image. This is caused by the shadow of the secondary mirror.

In a refractor the image will show the effects of chromatic aberration. At the point of best focus the image will have a yellow tint with a purple halo. Inside this focus the disk will have a reddish-purple color at the edge. As you move outside the focus the central image will become a red disk. As you move further out from the focus a green-yellow edge will appear until a bluish flare forms over the central image.

In all types of telescopes the brightness of the image should be the same across the disk. If it is brighter at the edges than the center, or vice versa, the objective is showing spherical aberration. If the diffraction rings flip as you move through the focal point the system has astigmatism. If the image is fan-shaped coma is present.

A star test can show you if your optics are poorly made or designed. If you took the time to make the telescope you are testing and any of these aberrations show up, it is time to return to the optical shop and perform a bit of optical surgery on your patient. If these things show up in a commercially produced instrument and you cannot get rid of them no matter how much you collimate the optics or yell and scream, take the instrument back!

✳5✳

Eyepieces and Mountings

A telescope with good optics and a lousy eyepiece is
a lousy telescope . . . .
A telescope with good optics, a good eyepiece, and
a lousy mount is even worse!

Overheard at AstroFest 1988

Many amateurs spend large amounts of money making sure that the telescope they use has the finest optics and the sturdiest mount available. Then they go out and buy bargain basement eyepieces and wonder why the images they get are fuzzy and out of focus! The telescopes used by both professional and amateur astronomers are made up of two interrelated optical systems. The primary optical train consists of the objective and any secondary mirrors involved in passing the image to the eye. The other optical system is the eyepiece. Most beginners overlook the importance of the eyepiece.

EYEPIECES

The eyepiece is a collection of lenses that magnify the image formed by the telescope's objective. It is made up of two main lenses: the field lens that faces the telescope objective and the eye lens that lies nearest the eye.

Eyepieces come in a variety of types, shapes, and sizes (Fig. 5-1). An eyepiece is usually identified by its focal length and a letter designating its type. For example, an 18mm Ke is a kellner lens with an 18mm focal length; a 6mm Or is an orthoscopic lens with a 6mm focal length.

Fig. 5-1. Eyepieces. (Courtesy Meade Instruments Corporation.)

The focal length of the eyepiece determines two important factors. First and most important it determines the size of the exit pupil of the eyepiece. By dividing the eyepiece's focal length by the focal ratio of the telescope you get the size of the exit pupil for that particular eyepiece. For example, if you want to know the size of the exit pupil of a 10mm eyepiece when it is used with a Meade 4-inch schmidt-cassegrain (SCT) telescope divide the focal length of the eyepiece by the focal ratio of the telescope and you will get an exit pupil of 1mm (10mm/10). The lens will have the same size exit pupil when used with any telescope with a focal ratio of f/10. If you put that same lens in a telescope with a focal ratio of f/5, the exit pupil becomes 2mm (10mm/5).

Secondly, the focal length of the eyepiece tells you the amount of magnification the eyepiece will impart on the image that passes through it. To obtain this figure the focal length, not ratio, of the telescope is divided by the focal length of the lens with both distances expressed in the same terms, either millimeters (mm) or inches (in). The 10mm eyepiece in the Meade SCT magnifies the image 100 times (1000mm/10mm). This is written as $100\times$ or $\times 100$.

Unfortunately, when most people think of a telescope they think of its magnifying power. As often as not, the first words out of a person's mouth when a telescope comes into view is something like "what power is that thing?" It doesn't do any good to magnify an image to the point where it becomes unclear and blurred. This wastes the real power of the telescope—its ability to see small details.

The eyepiece should make full use of the resolving power of the telescope. Remember that when the magnification is increased the field of view is decreased. Increased magnification not only increases the apparent size of the image, but it also increases the apparent speed of the Earth's rotation causing the image to move faster across the field of view. Increased magnification also magnifies each little tremor that passes through the optical train so that a puff of wind can make the image shake and jump until it becomes impossible to see anything of interest.

Eyepiece Aberrations

Because light must travel through the lenses that make up an eyepiece, the eyepiece is subject to the same aberrations as a telescope's objective. An eyepiece that causes the image of a star to flare as you shift your center of vision suffers from spherical aberration. Caused by the uneven focus of light, this aberration makes it difficult to focus. By their very nature lenses are subject to chromatic aberration, that annoying habit of adding a rainbow halo of color to a bright object. Distortion in a lens causes a straight line to curve as it approaches the edges of the field. *Field curvature* creates a lack of sharpness throughout the field of view with the center being in sharp focus while the edges are fuzzy and indistinct. Astigmatism in an eyepiece will cause the image of a star to appear as a cross or square the further away it is from the center of the field. Finally, because some eyepieces are made up of a number of optical surfaces the image may reflect off them as it passes through the eyepiece and cause ghost images to form. Tracking the source of an aberration can be difficult because both the eyepiece and the objective may suffer from the same set of defects.

Types of Eyepieces

The eyepieces that can be used on a telescope are determined by the size barrel the telescope can accept. The barrel of the eyepiece comes in three sizes. 24.5mm or .965-inch is the "Japan" standard size. Most of the small telescopes found in department and discount stores come with eyepieces of this size. The types of eyepieces available in this size are limited. The most popular size, known as the American standard, is the 1.25-inch barrel. Almost all eyepieces can be found in this size. In recent years the growing popularity of large-mirror, short-focal ratio dobsonians has lead to the development of 2-inch barrel eyepieces. These eyepieces are large and expensive and while the variety available is growing, the selection is not really that large as yet.

An adapter will let you use a smaller diameter barrel with a larger diameter focuser. You can use a 1.25-inch lens in both a .965-inch and a 2-inch focuser, but you cannot use a 2-inch barrel lens in a 1.25-inch focuser as yet. You must be careful, however, when trying an adapter between the 1.25-inch and .965-inch barrels. The focus of the 1.25-inch lens may be slightly longer than the travel of the focusing tube on some small refractors. Try to find out the amount of extra travel that the larger lens will require. A number of companies make and distribute star diagonals that are also adapters.

The huygens (H) eyepiece is the type supplied with many refractor telescopes sold in department stores. A cheap and easy eyepiece to produce, the huygens is best suited for the slow, longer focal ratios like f/15. While it produces an adequate field of about 30 degrees without ghost images, the huygens is plagued by other optical problems. The worst of these is its tendency toward severe spherical aberration when used with faster focal ratio systems. The huygens are also subject to distortion, curvature of field, and chromatic aberration. These aberrations can be minimized however, by using a long focal length system.

These eyepieces are difficult to find because they no longer come with the popular instruments produced in this country. Even the imported small scopes are moving away from this eyepiece type, but they can still be found if a person looks hard enough. If you have a telescope that comes with a huygens, don't throw the eyepiece away. They do have their uses as you will see. Instead wrap it carefully and put it in the back of the drawer.

Another lens that comes with today's small telescopes is the ramsden (R). Just as easy and cheap to produce as the huygens, the ramsden uses the same types of lenses as the huygens. The improvement made to the ramsden is these lenses now face each other, thus reducing some of the defects found in the huygens. Spherical aberration is greatly reduced. Chromatic aberration, especially at the edge of the field however, is still apparent.

The eyepiece also has its own set of problems. The eye relief is short requiring the observer to put his eye very close to the eye lens. The field of view is smaller than the huygens and ghost images are more noticeable. While the ramsden can be used with slightly faster optical systems, down to f/8, they should not be used with an instrument like a rich-field-telescope. The ramsden seems to be the eyepiece of choice for the importers. Most companies offer their scopes with a couple of ramsden eyepieces, usually a 12mm and a 4mm. Like the huygens, the ramsden will have its uses so don't discard it.

The kellner (K or Ke) is an improvement on both the huygens and the ramsden. Often called an achromatic ramsden, the kellner uses an achromatic doublet as an eye lens. Like the achromatic objective of a refractor, the qualities of the two types of glass that make up the doublet cancel out the aberrations caused by each other. The immediate advantage to this system is the reduction of chromatic aberration. While astigmatism and curvature of field are still apparent, they are greatly reduced. The large field of view, close to 50 degrees, and good eye relief make it an excellent eyepiece for low-power viewing. These eyepieces are subject to ghost images however, and to unfocused images near the edge of the field. Kellners are made by a number of companies and come in all barrel sizes. They come in focal lengths from 6mm to 40mm and range in price from $35 to $50, though they are often discounted.

In the 1970s, Edmund Scientific Co. commissioned Dr. David Rank to develop an improved version of the classic kellner eyepiece. Called the rank-kellner-eyepiece (RKE), it consists of an achromatic field lens and a single double-convex eye lens. The production is overseen by the company to exacting standards. The result is a lens with a very sharp field of 45 degrees. These lenses give the amateur on a budget a chance to obtain excellent quality eyepieces at a reasonable cost. The RKE eyepieces list at $40 direct from Edmund Scientific, and range in size from 8mm to 28mm in the American standard barrel size.

The orthoscopic eyepiece has long been considered state of the art and has been the lens of choice of serious amateurs for over two decades. It has a three-element field lens and a plano-convex eye lens that produces a field of 45 degrees with good eye relief and little chromatic aberration. It does, however, suffer from astigmatism and a strongly curved field.

Orthos are made by everyone! Many manufacturers contract out the construction of this type of lens to small Japanese firms. The result is that every company has a line of orthos carrying their name. Orthoscopic eyepieces come in all barrel sizes and range in focal length from 4mm to 25mm. Prices vary over a wide range starting at $40 for a private label .965-inch barrel lens and going all the way to $200 for a 2-inch barrel, although these lenses are frequently discounted. The popular 1.25-inch lens is priced between $50 and $80. A diligent check of the price lists in the major magazines will reveal a wide range of discounted prices.

The plossl is a recent design that uses two identical achromats to give good eye relief and a wide field of view. It is less troubled by astigmatism than the orthoscopic eyepiece, but has a slightly higher level of distortion. Plossls can be found in the American standard barrel size in focal lengths from 3.8mm to 45mm, and from 40mm to 56mm eyepieces in 2-inch barrels. Construction of these eyepieces has been turned over to foreign jobbers. As a result many private label plossls exist. Prices range from $60 to almost $300 for the big 2-inch sizes.

The erfle is the eyepiece preferred by the observer who wants a wide field with good definition. The lens placement varies among manufacturers, but usually includes an achromat for the field lens and eye lens, which are separated by a third lens, most often a double-convex lens. A popular lens for deep-sky observing, the erfle can be found in the 1.25-inch barrel with a size range from 20mm to 40mm. Prices for a good erfle are right around $100.

The konig is a variation of the erfle. It is a bit simpler in design but produces an excellent image and field definition.

State-of-the-art eyepieces for the late 1980s are the nagler and wide-field eyepieces designed and made by Tele-Vue Optics. With its wide fields and excellent definition, the nagler is becoming increasingly popular with deep-sky observers. This eyepiece can be found in sizes ranging from 4.8mm to 20mm for both 1.25-inch and 2-inch barrels. It ranges in price from $190 to $450. A more recent design, the wide-field eyepiece rivals the erfle in field size, definition, and eye relief. Sizes range from 15mm to 24mm for the 1.25-inch barrel and from 32mm to 40mm for the 2-inch barrels. Prices are from $165 to $205 for the smaller size and from $290 to $360 for the larger.

Finally, there is the barlow lens. Not really an eyepiece at all (Fig. 5-2), the barlow acts to increase the focal length of the eyepiece with which it is used. The barlow lens increases the power of the eyepiece two to three times. Most eyepieces are restricted to a fixed amount of magnification, but some can be changed by changing the distance between the field lens and the barlow lens. Using a barlow lens with an 18mm eyepiece

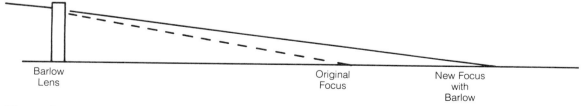

Fig. 5-2. The barlow lens acts to lengthen the focal length of your objective allowing you to get a higher magnification from your eyepieces.

gives you the equivalent of a 9mm with a 2× barlow or a 6mm with a 3 × barlow. The barlow effectively multiplies the number of lenses in your collection by two. If you have a 12mm lens and a 2× barlow, you don't have to buy a 6mm lens because you already have it when you combine the 12mm with the barlow.

Another advantage of the barlow is that it retains or increases the eye relief of the original lens, a boon to those who wear glasses. Barlow lenses are another item that seems to be made by everyone, for every barrel size. One important thing to remember about the barlow is that the field lens should be adequate in size or the image brightness will suffer drastically. Prices range from $50 to $120.

Selecting an Eyepiece

The best eyepiece for any particular observational need meets two simple qualifications: it fits your observing needs and your budget. If your observational interests lie with planetary and lunar observations you will want an eyepiece that gives very sharp, clean images. Orthoscopics, plossls, and RKEs fit this bill of fare very nicely. If you want sweeping views of deep-sky objects like star clusters, a lens with a wide, flat field of view like an erfle or a konig is more appropriate.

The size of the eyepiece is another thing to consider. Remember that the important feature of the eyepiece is the size of its exit pupil, not its magnification. Use the information in Table 5-1 as a guide for selecting a range of eyepieces.

If, for example, you use an 80mm refractor telescope with a focal length of 910mm, what types of eyepieces should you have in your observing kit? The telescope's focal ratio is slightly over f/11, f/11.4 to be exact. To keep the math simple let's stick to f/11. According to Table 5-1 an eyepiece for general viewing should have an exit pupil of 3mm or 4mm. In order to get this exit pupil, you will need an eyepiece with a focal length of 33mm. A 33mm eyepiece is a awful lot of glass and is going to cost a pretty penny. A 32mm plossl will fit your needs, but they can be expensive. Konigs are popular and less expensive. An erfle would be ideal but, for reasons unknown to me, not many places carry them. Let's say you decide to go with a 32mm erfle and after a bit of hunting you track one down for about $60.

Table 5-1. Exit Pupils

Exit Pupil Size (mm)	Object Type Viewed
5 - 7	Galaxies
	Nebulae
3-4	General viewing
	Clusters
	Planetary nebulae
2	Moon and planets
	Double stars
1	Maximum for moon and planets
	Close doubles
0.5	Closest doubles on best nights

Source: Jeffrey L. Hunt, Cosmic Connections, Inc.

For viewing the Moon, planets, and double stars the table recommends an exit pupil of 2mm. With our telescope that would work out to a 22m focal length eyepiece. You have a variety of eyepieces that will give this exit pupil. A 22mm plossl or a 21mm RKE will both be adequate for your needs. You decide to go for the quality of the plossl and spend $110. Total so far: $170.

When seeing is good and you want a close look at the belts on Jupiter you will need an exit pupil of 1mm or an 11mm focal length eyepiece. You have a 12mm orthoscopic, RKE, or private label plossl from which to choose. This time you decide to go with "the new kid on the block" and choose the 12mm RKE for $40. Total spent on eyepieces: $210.

For those nights when the atmosphere is calm and the seeing perfect, a .5 exit pupil could be used to split doubles right at the limit of the telescope. A 5mm or 6mm focal length lens is what you are looking for and, again, there are many choices. You decide that for an eyepiece of this focal length an ortho is ideal. You spend another $70 for a good 6mm lens. This brings your final total to $280.

You now have four eyepieces ranging in size from 6mm to 32mm, with exit pupils from .5mm to 3mm. What magnification range do you have? The 32mm erfle gives you a magnification of ×28, the 22m plossl gives ×42, the 12mm RKE ×76, and the 6mm gives you a magnification of ×152. The magnification per inch of aperture ranges from ×9.5 per inch for the 32mm erfle to ×50 per inch for the 6mm ortho. These figures fall perfectly within the optimum for the telescope's best performance.

TELESCOPE MOUNTS

Exit pupil size and magnification mean nothing if the object you are looking at is constantly jumping around the field of view. Poor mountings have contributed to more than one amateur astronomer trading in the telescope for a seat in front of the wide screen television.

A good, solid mount is an essential piece of equipment for the amateur. The mount should meet these two basic requirements: it should provide solid support for the weight of the telescope; and it should be able to move easily along both the horizontal and vertical axes.

Support is the main job of the telescope mount. It must provide a stable base that will dampen any vibrations that travel through the system. A mount won't eliminate vibrations, but it should minimize them to a point where they disappear quickly. One of the primary contributors to the poor image of the small telescope is the shaky mount. The optics of a small telescope may be very good, but if the mount is flimsy the optics are almost useless.

Altazimuth Mounts

The type of mount responsible for all this trouble is a cheap variation of the altazimuth mount. This mount usually consists of a pillar and claw. The telescope is balanced between two points on the claw. This type of mount is so unsteady that many come with a bracing bar that further complicates things. When a telescope is mounted on a tall, unbraced, and rickety tripod, it is a wonder anything is seen at all.

Altazimuth mounts suffer from another shortcoming, they make it difficult to follow the course of the star under observation. Because of the way stars move across the sky the telescope needs to track along two axes. Many refractor telescopes come with "slow motion" cables that make tracking easier, though tracking with an altazimuth mount is still difficult.

This is not to say that all altazimuth mounts are unstable, they aren't. I did all of the observational work for this book with an 80mm refractor telescope on an altazimuth mount (Fig. 5-3). The mount is stable and offers excellent support. The concept behind the altazimuth mount is simple: provide a simple, easily constructed base for the telescope. In the nineteenth century many telescopes were carried on altazimuth mounts. Many of the pioneering discoveries made by amateurs were made with altazimuth-mounted instruments.

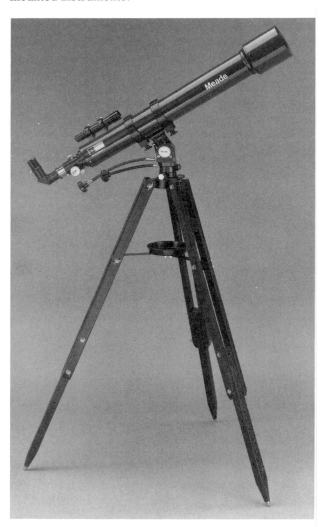

Fig. 5-3. An 80mm refractor on an altazimuth mount. (Courtesy Meade Instruments Corporation.)

A popular version of the altazimuth is the dobsonian mount that is used on many large reflectors (Fig. 5-4). The dobsonian is designed to provide a stable platform that is still easy to move when tracking celestial objects. The mount carries the weight of the main portion of the telescope—the primary mirror and its supporting cell—inboard or between its supporting columns. This makes the mount very stable and solid.

Another advantage of the dobsonian mount is its ability to move along the two axes simultaneously with very little effort. You can push the telescope easily, owing to the use of teflon for bearing material, and when you stop, the mount stops.

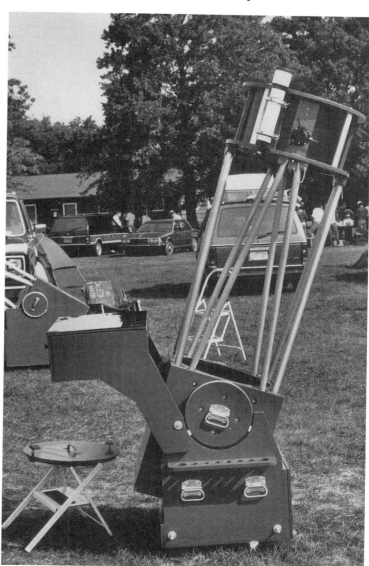

Fig. 5-4. Another form of altazimuth mounting is the dobsonian mount.

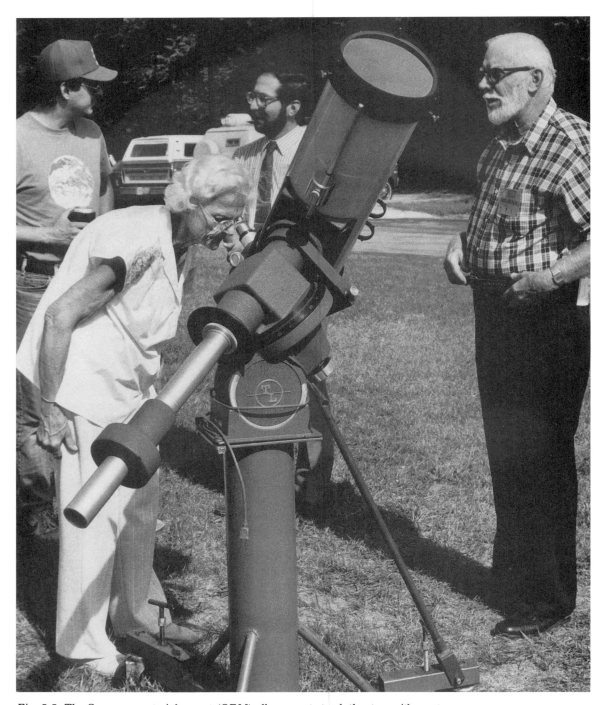

Fig. 5-5. *The German equatorial mount (GEM) allows you to track the stars with great ease.*

Equatorial Mounts

The best mounting for any telescope, no matter how small, is the equatorial mount. Invented by the German scientist Joseph Von Fraunhofer, the equatorial mount is designed to follow the movement of a celestial object with a minimum of effort. The modern equatorial has a variety of shapes, each one with its own distinct advantages and disadvantages. The most popular of these is a direct descendent of Fraunhofer's mount and is called the German equatorial mount or GEM (Fig. 5-5).

The altazimuth mount tracks a star by moving through two perpendicular axes. This means that for every degree of movement made by the star, the telescope and mount must make two movements. The equatorial mount simplifies this by making one of its axles parallel to an axis. This is the polar axis of the mount. It extends from the mount toward the north celestial pole. With this system one of the two axes is stationary so the mount has to swing the telescope through only one axis—the right ascension axis. This setup also allows for the addition of a motor or clock drive, to make the movement through this axis automatic. The equatorial mount also allows for the use of setting circles on each axis to index star and object positions (Fig. 5-6).

Another popular mount is an equatorial variation called the fork mount (Fig. 5-7). Found on the SCT telescopes, the fork shrinks the equatorial mount to a size that will fit in a small box. The fork mount points the body of the mount toward the pole until the "tines" of the fork are parallel to the polar axis. The telescope is then cradled inboard between the tines for a vary stable configuration.

Aligning an Equatorial

One of the few advantages of the altazimuth mount is the fact that you need not worry about where or how to set it up. Just step outside, plant the tripod, and you are ready to go. With an equatorial mount it isn't that easy. Unless you have the mount set on a permanent pier, you will have to align the mount each time you use the telescope.

If you don't have your mount fitted with a clock drive you may just need a rough alignment. To do this, simply sight along the mount in the direction of Polaris. Depending on your skill the mount should be within 5 to 10 degrees of the pole. If you only plan a short viewing session this is adequate.

For a better alignment, first use the rough alignment technique and then clamp the telescope at 90 degrees declination. Move the entire mount until Polaris is nearly centered in the telescope's finder. This should improve the accuracy to within 2 to 3 degrees of the pole.

Many of the commercially available mounts come with a handy feature called a polar alignment reticle. This little gadget is designed to show you exactly where the pole is in relation to Polaris. Mounted in the finder of the SCTs, you simply give the telescope a rough alignment by eye, line Polaris up in the finder, adjust the finder so that the pole is at the proper mark on the reticle, and you are ready to go. Many companies also offer this feature with a GEM mount. All you do is sight through the polar axis because the mount has a finder scope built in, complete with a polar alignment reticle!

Fig. 5-6. If you attach a motor drive, also called a clock drive, to a GEM the telescope follows the stars automatically. Setting circles also allow the observer to locate objects quickly.

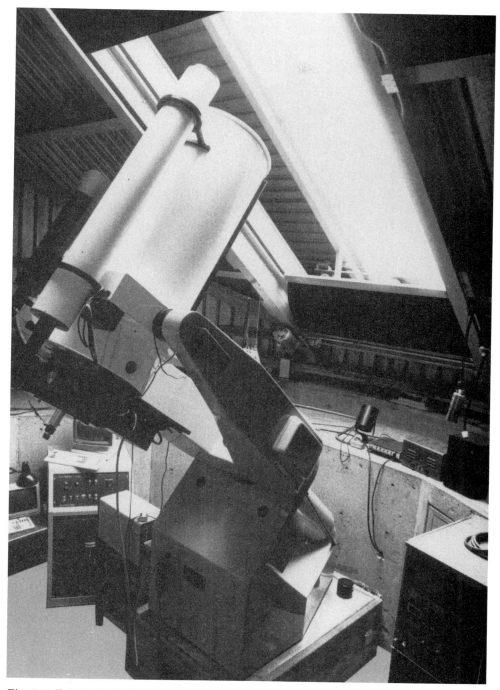

Fig. 5-7. Today's SCTs have a mounting in common with large observatory instruments like this. They both use a form of equatorial mounting called the fork mount. (Courtesy Adler Planetarium.)

✶6✶

Finding Your Way Around the Sky

I can see it with my naked eye,
but I can't find it with the telescope!

Overheard at a Star Party, 1988

The next time you find yourself away from the lights of a city take a minute and give the night sky a quick glance. Make it quick because looking at the night sky undimmed by light pollution is like gazing at the face of Medusa. It may be entrancing but it will turn you to stone! The seemingly countless number of stars has dazzled many budding amateurs and sent them scurrying for the shelter of the big screen television. How can anyone find anything in a star-choked sky like that, they ask. Tracking down a star cluster in the apparent maze of stars is not that difficult. All it takes is a little inside information and a bit of practice.

COORDINATES ON THE SKY

One method of finding your way around the sky is to assign coordinates to celestial objects. This way you will know where any object is at any given time. Once you decide to assign coordinates to celestial objects you will need to determine the type of coordinate system you will use.

Right Ascension and Declination

In the fifteenth century, an enterprising Italian inventor and mapmaker named Toscanelli dal Pozzo was intrigued by the inability of navigators to find their way around the seas after they lost sight of land. He reasoned that if a grid of equally spaced lines was placed over a map it would be easier for a ship to know where it was and where it was going. The grid lines he placed on his map went on to become the lines of longitude and latitude that we use today. It took another 300 years for this grid to make it from the Earth to the sky.

Called the *equatorial coordinate system* (Fig. 6-1), it takes the Earth's equator and projects it onto the dome of the sky to get what is called the *celestial equator*. At an angle of 23.5 degrees to the celestial equator is the *ecliptic* or apparent path the Sun follows across the background of stars in the course of a year. In the same manner, if you project the north and south poles of the Earth into space you get the *north* (NCP) (or *south* (SCP)) *celestial pole*. With these points of reference the final step is to project the lines of longitude and latitude onto the dome of the sky.

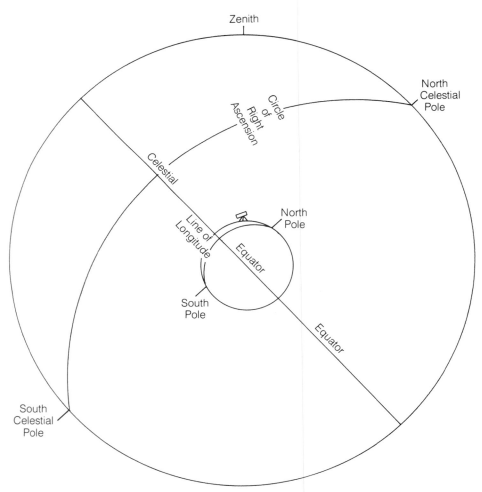

Fig. 6-1. *The equatorial coordinate system is the extension of the Earth's own system into space.*

The east-west position measurement given by longitude on Earth now becomes *right ascension* (RA). It is measured in hours (h), minutes (m or '), and seconds (s or ") starting at 00 hours, 00 minutes and 00 seconds at the Equinox and increasing as you move eastward along the celestial equator from the point of the Vernal Equinox. RA is the difference between the object's time of transit and the time the Vernal Equinox transits the meridian.

The *declination* of an object is the distance, measured in degrees (°) and minutes (m or '), of the object either north or south of the celestial equator. A plus sign (+) indicates the object is north of the equator. A minus sign (−) indicates it is south of the equator. A declination of + 56 degrees tells you the object is 56 degrees north of the celestial equator; − 10 degrees puts the object 10 degrees south of the equator. The altitude of Polaris above your northern horizon is equal to your latitude. Straight overhead is what is called your latitude line (Fig. 6-2). If you live at 41.5 degrees north latitude the declination line that runs through your zenith will be 41.5 degrees.

Fig. 6-2. Stars near the north celestial pole never set. These circumpolar constellations are visible all year around. (Courtesy Walter Piorkowski, ©1988.)

The highest point in the sky that an object can reach is its *co-latitude*. This is particularly useful when trying to locate objects with a declination far below the equator. Add your latitude to the object's declination to find the co-latitude. Chicago has a latitude of 42 degrees north. If I want to find out how far an object at declination −33 degrees is above the southern horizon, I add 42 + (−33). That tells me the object is 9 degrees above the southern horizon. If the co-latitude of the object is zero or less than zero, it will never rise above your horizon.

Horizon Coordinates

A somewhat simpler set of coordinates can place the observer at the center of things. Using *horizon coordinates* the location of a celestial object is determined by its relation to the observer and the observer's horizon. This system has been used for centuries and is still used primarily for navigation.

To get the basics of the system (Fig. 6-3), go outside and face north. The point directly above your head is called *zenith*. The point exactly opposite the zenith, or directly below you at the center of the Earth, is called *nadir*. Directly in front of you is the *horizon*, or the line where the sky seems to join the Earth. This line travels in a great circle all the way around you. The north point on the horizon is the point directly under the north celestial pole. The line that stretches from this point on the horizon through the zenith and ends at the south point is the *meridian*. When an object crosses the meridian it is at its highest altitude above the horizon and said to be in *transit*.

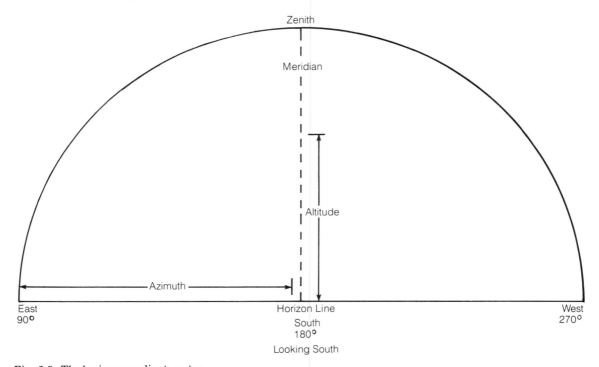

Fig. 6-3. The horizon coordinate system.

The horizon system is excellent for an observer trying to describe where an object is on the celestial sphere. Any object's position can be described by its *altitude* and its *azimuth*. Altitude is measured in degrees starting from the horizon and moving toward the zenith. An object halfway up into the sky would have an altitude of 45 degrees. If the object was directly overhead it would have an altitude of 90 degrees. Azimuth is measured in degrees from the north point on the horizon toward the east. The north point is 000 degrees, the point due east is 090 degrees, the south point is 180 degrees, and the west point is 270 degrees. An object above the horizon has the same azimuth as the point on the horizon directly below it. So a star directly over the west point on the horizon would have an azimuth of 270 degrees.

TIME

We keep track of time by using a basic unit of measure called the *mean solar day*. It is based on the time between transits of the Sun. In order to simplify everyday activities, the solar day we use is based on a Sun that moves against the background of stars at a constant rate of 24 hours between each transit. Unfortunately this is not an accurate rate because the Earth's orbit is not circular but elliptical. As the Earth moves in its orbit around the Sun, the Sun appears to move at different rates and does not transit at the same time everywhere it is visible.

To lessen confusion *local time* and standard time zones were invented. With this system the sun transits the meridian at noon no matter where you are. As you pass from one time zone to the next you change your watch by one hour depending on whether you are traveling toward or away from the central time keeping meridian, the *prime meridian*.

Astronomers selected the location for this meridian in much the same way they selected the point on the celestial sphere that marks the 00h00m00s coordinate in right ascension. They said, "This is as good a place as any, besides there is an observatory already here so we will not have to build one." So the prime meridian, the meridian from which all time is measured on Earth, happens to run through the Royal Observatory in Greenwich, England. Local time in Greenwich is also the basis for *universal time* (UT), which is measured on a 24-hour clock and used as a standard method of keeping time by both amateur and professional astronomers.

As you move west from Greenwich you begin to fall behind UT and must add an hour for each time zone you are west of Greenwich. Table 6-1 shows you how to add to or subtract from local time to get UT. If you live in Chicago like me, for example, you are six time zones away from Greenwich so you must add six hours to your time to get UT. This means that 1500 hours, or 3 P.M. Central Standard Time, in Chicago becomes 2100 hours UT. If an event is to begin at 0400 hours UT on June 24, subtract six hours to get 10:00 P.M. Central Standard Time. Remember that the date changes when you pass 0000 hours so you lose one day making it the 23rd instead of the 24th. Also remember to adjust for daylight savings time.

Sidereal time is another kind of time used in astronomy. Sidereal time, simply put, is the right ascension of objects that are on the meridian at that moment. For example, the sidereal time for midnight, July 15, is 19 hours and 30 minutes. Now you know that any object with a right ascension of 19h30m will be on the meridian at that instant. Sidereal time makes it easy to use the coordinate systems we looked at in the beginning of this

Table 6-1. UT/Local Time Conversion

UT = Local Time + Cor.

LOCAL TIME IN	CORRECTION
Eastern Standard	+ 5 hours
Eastern Daylight	+ 4 hours
Central Standard	+ 6 hours
Central Daylight	+ 5 hours
Mountain Standard	+ 7 hours
Mountain Daylight	+ 6 hours
Pacific Standard	+ 8 hours
Pacific Daylight	+ 7 hours

Example: If you live in Chicago you are in the Central Time zone. To convert 6 P.M. CST to UT, first convert your local time to 24-hour time by adding 12 hours to P.M. times (leave A.M. alone). So 6 P.M. CST becomes 1800 hours (sound familiar vets?). UT = local time (1800) + cor (+6 hours). UT = 1800 + 6. UT = 2400 (or 00).

chapter to actually find something in the sky. Table 6-2 gives the sidereal time of objects transiting the meridian for any hour of any day of the year.

If you know the sidereal time and the right ascension of any celestial object you can find out if that object is favorably placed for observation. This is because stars on the equator (Fig. 6-4) rise about six hours before they transit and set about six hours after transit. If you subtract the right ascension of the object from the sidereal time you find its *hour angle*.

The hour angle tells you how far the object is from transit or how far it has moved since transit. If the sidereal time is smaller than the right ascension, then the object is east of the meridian. It has not yet reached the meridian and its best placement for viewing. If the sidereal time is greater than the right ascension then the object is past the meridian and is on its way toward the horizon and setting. If the sidereal time and right ascension are equal, the object is on the meridian and it's time to get the telescope out! You can use Table 6-2 to tell you when any object for which you know the RA will transit.

ROAD MAPS TO THE STARS

Every amateur needs a good star atlas. A star atlas is nothing more than a detailed map of the sky broken down into manageable segments. They are different from the constellation guides you used earlier in two basic ways. First, they show what the constellations look like when viewed in total. A small guide like *Olcott's Field Guide to the Skies* is excellent for the beginner, but it doesn't give you a feeling for the area of the sky that it actually covers. Don't get me wrong, I really love these little field guides. I still have and use my Olcott's after 25 years. It's just that if you are doing serious observing you need a more in-depth atlas.

Table 6-2. Sidereal Time

Time	8 P.M.	9 P.M.	10 P.M.	11 P.M.	12 A.M.	1 A.M.	2 A.M.	3 A.M.	4 A.M.
Date			Right Ascension on Meridian						
Jan 5	03h	04h	05h	06h	07h	08h	09h	10h	11h
21	04h	05h	06h	07h	08h	09h	10h	11h	12h
Feb 5	05h	06h	07h	08h	09h	10h	11h	12h	13h
20	06h	07h	08h	09h	10h	11h	12h	13h	14h
Mar 7	07h	08h	09h	10h	11h	12h	13h	14h	15h
22	08h	09h	10h	11h	12h	13h	14h	15h	16h
APR 6	09h	10h	11h	12h	13h	14h	15h	16h	17h
22	10h	11h	12h	13h	14h	15h	16h	17h	18h
May 7	11h	12h	13h	14h	15h	16h	17h	18h	19h
22	12h	13h	14h	15h	16h	17h	18h	19h	20h
Jun 6	13h	14h	15h	16h	17h	18h	19h	20h	21h
22	14h	15h	16h	17h	18h	19h	20h	21h	22h
Jul 7	15h	16h	17h	18h	19h	20h	21h	22h	23h
22	16h	17h	18h	19h	20h	21h	22h	23h	00h
Aug 6	17h	18h	19h	20h	21h	22h	23h	00h	01h
22	18h	19h	20h	21h	22h	23h	00h	01h	02h
Sep 6	19h	20h	21h	22h	23h	00h	01h	02h	03h
21	20h	21h	22h	23h	00h	01h	02h	03h	04h
Oct 6	21h	22h	23h	00h	01h	02h	03h	04h	05h
21	22h	23h	00h	01h	02h	03h	04h	05h	06h
Nov 6	23h	00h	01h	02h	03h	04h	05h	06h	07h
21	00h	01h	02h	03h	04h	05h	06h	07h	08h
Dec 6	01h	02h	03h	04h	05h	06h	07h	08h	09h
21	02h	03h	04h	05h	06h	07h	08h	09h	10h

That brings us to the second major advantage of a good star atlas—its detail. Most of the star guides show the stars down to the naked-eye limit or magnitude 6. When you find yourself at the eyepiece of a telescope or a good pair of binoculars you are seeing stars well past that limit. Depending on the size of your telescope, you could be trying to pick your way through a field of tenth- or eleventh-magnitude stars, which can sometimes be confusing. Even if you only use an accurate finderscope you are going to be able to see to magnitude 9 or better, so you need an accurate guide to what is out there. It is frustrating, and a bit embarrassing to spend time at the telescope and never be quite sure of just what you are seeing. With a good star atlas you will always know just where you are looking and what you are viewing.

Star atlases come in a number of sizes and types. They range from simple, easy to use atlases like *Edmund's Mag 6 Star Atlas* to the newly published and extremely detailed *Uranometria 2000*, which shows stars down to magnitude 9.5. Falling nearer to

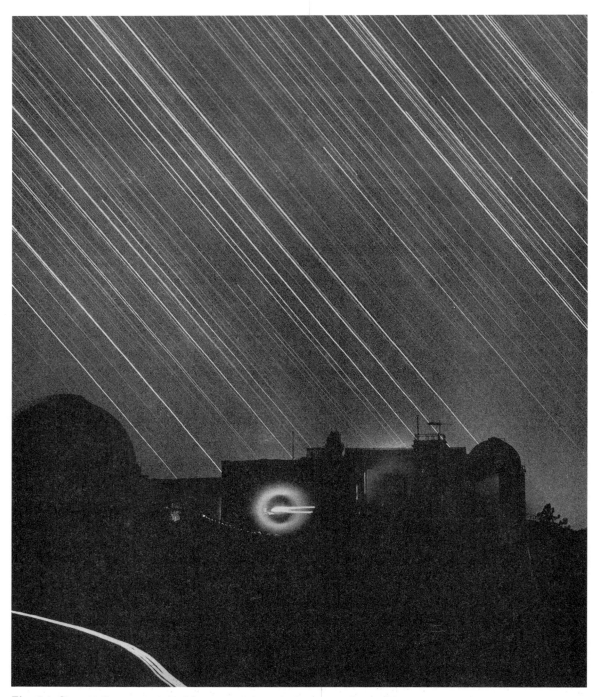

Fig. 6-4. *Stars at the equator take 6 hours to go from the horizon to the meridian and another 6 hours to go from the meridian to the horizon again. (Courtesy Lick Observatory.)*

the Edmund atlas is probably the best star atlas for a beginner. It is the classic *Norton's Star Atlas and Reference Guide*. Originally published at the end of the nineteenth century, Norton's has been a stand-by for amateurs for almost 100 years. One of the wonderful things about Norton's is that the maps are only part of its content. It is also packed with such things as definitions of astronomical terms, information on stars and star structure, and maps of the Moon and Mars. Its 16 maps cover the sky from the north to the south celestial pole. Although it only goes down to magnitude 6, there are many fainter stars listed, especially variable and double stars. Most atlases show magnitudes by varying the size of the star images. Norton's makes each magnitude a distinct and easily recognizable image so it is easy to tell the difference between a third- and fourth-magnitude star.

This discussion of magnitude representation brings to mind something that you must remember when using most star atlases. Many atlases on the market use the ''lump'' system of magnitude representation. By this I mean that they take all the stars within a certain magnitude range, say between magnitude 4.0 and 4.9, and list them as magnitude 4 using the same symbol. I can understand why they do this. It would be almost impossible to distinguish little dots showing stars of magnitude 4.3 and 4.6. Long time users are aware of this practice and do not have a problem using an atlas set up this way, but it can be confusing for beginners.

Other star atlases that the beginner might consider using are *Sky Atlas 2000* and *Uranometria 2000*. Two more are the venerable *Smithsonian Astrophysical Observatory (SAO) Atlas* and the *AAVSO Star Atlas*, both of which go to magnitude 9.0. *Atlas Stellarum* is published in two volumes, one for stars from the north celestial pole to declination -25 degrees and another for stars from declination -15 degrees to the south celestial pole. It has a limiting magnitude of 14. These atlases, however, are expensive and hard to find. I have looked them all over and find them a bit hard to use because there is nothing to indicate the star at which you are looking without resorting to the use of overlays and a catalog to pinpoint the star for which you are hunting. Consider trying these atlases after you've become adept at using an atlas like *Sky Atlas 2000*. Trying to use them at the beginning of your astronomical apprenticeship may just drive you inside to the big screen and reruns of *My Mother the Car*!

Sky Atlas 2000 was designed and plotted by Wil Tirion and is concise and easy to use. It comes in three editions: the field edition with white stars on a black background; the desk edition with black stars on a white background; and the deluxe color edition with the Milky Way and deep-sky objects color coded so there is no way to confuse a faint galaxy with a globular cluster. Tirion went to great pains to make *Sky Atlas* a usable atlas. He has succeeded marvelously. The atlas breaks the sky into 26 manageable sections. Each star is listed with its Greek letter designation or its Flamsteed number along with its common name. Alpha Lyra, for example, is listed as Vega, Alpha Lyra, and 3 Lyra so you know what you are looking at on the chart.

Uranometria 2000 is a two-volume set. Volume 1 covers the sky from the NCP to declination -6 degrees. Volume 2 covers the sky from declination $+6$ degrees to the SCP. Each volume contains 259 charts with a wealth of detail for the deep-sky observer. The publication of this atlas has not been without some controversy, however, in the amateur astronomical community. The primary complaint has been that the plotted positions of stars and deep-sky objects were taken from source material that was occasionally flawed and that those flaws found their way into the present atlas. I use a copy of *Uranometria* and do not find any of these flaws to be really disastrous.

Regardless of what atlas you decide to use take a little time to get to know it. Learn what symbol represents each deep-sky object and how those objects are named. One drawback of Norton's is that all the deep-sky objects—clusters, nebulae, and galaxies—carry the same symbol and are named under the old Herschelian system that divides deep-sky objects into eight subclasses. Using this system, the magnificent Double Cluster in the constellation Perseus is labeled H.VI 33 and H.VI 34. *Sky Atlas 2000* clearly labels it as the Double Cluster with its NGC designation of NGC 869 and NGC 884 next to it.

FINDING YOUR TARGET

Picking an individual star, or deep-sky object out of a star-filled sky has given more than one amateur fits. At first glance it would seem impossible to pick one's way through the vast starfields of Cygnus to locate a small open cluster of stars or a variable star near minima. It may seem impossible but it is not.

There are two basic ways for you to find objects in the clutter of the sky. You can dial up your target by determining the object's coordinates and adjusting your telescope's setting circles to those numbers. The target is there, yes? There was a time, not all that long ago, when the only telescopes that could use this method were those with an equatorial mount. Setting circles are nothing more than indices that give a numeric representation of where the telescope is pointed. Today, however in the age of microchips and mini-computers even the lowly altazimuth mount can be fitted with devices to make pointing the telescope easier. A hand calculator and a couple of protractors can enable you to find faint objects using the same system with any type of mount.

Remember our discussion of coordinate systems at the beginning of this chapter? This is where that stuff gets put to use. A star can be located if you know its declination and right ascension. Of those two figures, the RA is the most important to use with setting circles. To use the setting circles on an equatorial mount set the telescope on a star for which you know the coordinates (Fig. 6-5). Adjust the RA circle so it agrees with the published coordinates. Your telescope is now in tune with the sky. If you have a clock drive turn it on. The setting circles will remain correct for the rest of your observing session. If you don't have a drive, don't worry. You will have to reset your circles periodically throughout the session but, as we have seen, this is not a difficult procedure.

Using circles on an altazimuth mount, like a dobsonian, is a bit more complicated, but only until you get used to it. Using a simple set of equations you can convert the equatorial coordinates of any object to coordinates for the objects azimuth and altitude. On an equatorial mount the setting circles are usually built in, but with an altazimuth mount you are going to have to make your own. There are a number of ways to do this. The simplest is to use a drafting protractor for the altitude circle and scribe out an azimuth circle on the telescope ground board. With these additions and a hand calculator or computer you can easily convert equatorial coordinates to altitude and azimuth to make a setting circle system that really works. A number of recent articles describe the computer programs for these conversions. I have listed them for you in the Bibliography.

Fig. 6-5. The setting circle allows you to use the equatorial coordinate system to dial the position of a star and find it quickly.

Many amateurs taken with technology say that using setting circles and computers to find objects in the sky allows you to see many more objects during an observing session. But is that what amateur astronomy is all about? Is it nothing more than a numbers game where you try to find more and fainter galaxies than the other amateurs you know? For some of us this is true. I know of one amateur who preens at the fact that his computer system allows him to see over 30 deep-sky objects in just one hour. Pretty good, eh? Let's see 30 objects in 60 minutes . . . that comes out to two minutes per object. What! Just two minutes? Isn't that giving the myriad of detail locked in the Rosetta nebula the short end of the stick? Many amateurs feel that it isn't, but then again many of us feel that you should take your time with the sky and its wonders. There is no need to race through the constellations. Relax, enjoy the view!

Please, don't get me wrong. I am a great believer in using technology for all it's worth and then some, just ask my wife. What I am saying is that there is no need to take that technology and run a race with it. The technology of the microchip was developed so that we could be relieved of the drudgery and boredom that certain things bring, in this case the computations for determining coordinates. I personally feel that this technology is best used when it allows you to take the time to really look at an object like the Ring nebula or the Whirlpool galaxy, not to ring them up as points on a cosmic pinball game. All this brings us to the other way of finding objects in the night sky—star hopping. Star hopping is one way for an observer to really get to know the night sky. It can be confusing and even frustrating at times, but I think it is worth the effort.

The basis for star hopping is simple. You know the location of object A, which is fairly bright and obvious, and you know where object B is, which is rather faint and hard to see. What you want to do is get from A to B as simply and as easily as possible. I use this method a lot in my study of variable stars. I start with a fairly bright star and work my way over to the variable star field I want to study. Your target doesn't have to be a variable star field, it could be a nebula or some other deep-sky object.

What star hopping allows you to do is become even more familiar with the night sky that crosses your telescope field. It also lets you get to know your instrument better. You learn the little foibles that hide inside a refractor or reflector telescope. It allows you to learn your limits and the limits of your equipment and in the process helps you push those limits back a little. So let's take a few minutes and see just how star hopping works.

The first things you have to know are the field sizes of both the finderscope and eyepiece you will be using. Sounds pretty scary doesn't it? Well it's really quite simple. Take your telescope out one night when you can see the Big Dipper, which is any night because you can always see this constellation. Aim the finder at the two pointer stars (Fig. 6-6). Do they fit in the finder's field? If they do, your finder has a field of 5 degrees because they are 5 degrees apart. If you have a lot of room left over in the finder field check the bottom two stars in the bowl. Do they fit? If so, your finder has a field of 8 degrees because those stars are 8 degrees apart.

Most finders manufactured today have fields between 5 and 8 degrees. Using this method, I determined that the finder on my 80mm refractor has a field of almost 8 degrees because the bottom two bowl stars don't quite fit. I rounded the field off to 7 degrees, and this figure has served me well for a long time. Now you know your finder's field in degrees. So what good is that you might ask. I'll tell you, but first scurry inside and find a coathanger and grab your star atlas. Take a look at the introduction and find the scale of the atlas. If there is nothing like this listed measure the distance between the declination lines along the side of the charts. In my copy of Norton's, using my rule, I found that 1 degree equals 3.5mm so the scale is 1 degree of sky equals 3.5mm. Now

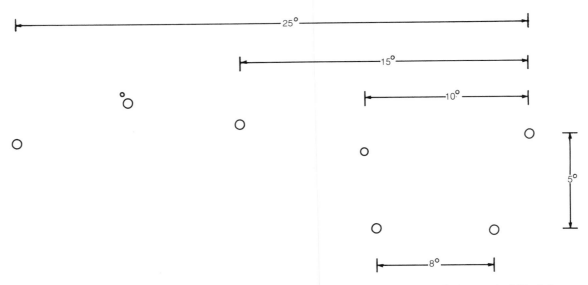

Fig. 6-6. The Big Dipper offers a ready made measuring stick to find your binoculars or finderscope's field of view.

take that coathanger, snip off the straight part at the bottom, and form a circle equal to your finder's field times the scale of your atlas. For my finder this comes out to 7 times 3.5mm or 24.5mm. Place a bit of tape over the rough edges so you don't cut yourself and you have an instrument that will tell you what you will be seeing through your finder. Lay this precision instrument on any part of your atlas and you will be able to see how much of the sky around a star will be visible to you through the finder.

Now you have to determine the field of your eyepiece, so go back outside again. This time select a star from your atlas that is on or near the celestial equator. A good star to use during the fall and winter months is the top star in Orion's belt, Delta (δ) Orion. In the spring and early summer use Gamma (γ) Virgo and during the late summer and early fall use Theta (θ) Aquila. Find the star and center it in the eyepiece. Notice how it moves across the field. Now position the star so that it moves across the field, as close to the center of the field as you can. Take your stopwatch, or a watch that keeps accurate time in seconds, and time the star as it moves across the field in minutes and seconds. Do this a couple of times and take the average of your timings. Multiply the result by 15 and you will have the field of the eyepiece in minutes of arc. This is because it takes a star on the celestial equator 15 minutes to move through one degree. Finally, divide the final result by 60 and you have your field in degrees.

For example, if it takes your target star 5 minutes to cross the field multiply 5 times 15. You will find that the field is 75 arcminutes across. Divide that by 60 because there are 60 arcminutes in one degree, and you will find that the field is 1.25 degrees across. Now go back inside and twist the other end of that precision coathanger into a circle 3.5mm times 1.25 across. The result is a circle about 4.3mm across. That is the area you will see in the eyepiece. Quite a bit smaller than the circle you made for the finder isn't it!

Now you are ready to star hop with the best of them! Let's try to find the Dumbbell nebula in Vulpecula using this method. Coathangers ready? We will be using chart 13 from Norton's so find the proper page and turn the book upside down. That's right, upside down! Remember your finder is a refractor and as such it gives you an inverted view of the sky. Turning the atlas upside down matches this view and makes it easier to follow along at the telescope.

Pull up a chair and get to work. First let's find Beta (β) Cygni, better known as Alberio. Place the larger of your rings so that it is centered in the circle. You will notice that a star labeled Alpha in Vulpecula is at almost 12 o'clock in your field. Moving around the circle you will see a star listed as 10 Vulpecula at about 2 o'clock. This is our next target. Now move your ring so 10 Vulpecula is centered. Ready? By the way, feel free to take notes as you go along, they will help once we get to the telescope.

With 10 Vulpecula centered you will see our next stop labeled 13 Vulpecula and located again at about 2 o'clock. Now move your ring to center this star. Notice that there is a group of three stars just above 13. Our final target, the Dumbbell nebula, is just above that center star. Take the smaller ring and center that on this star, labeled 14 Vulpecula. The circle covers both 14 and the Dumbbell, which is listed as M27. This way you know that if you can see 14, the Dumbbell isn't far away.

See, it's easy. We reached an object that is beyond our naked eye by hopping from Alberio to 10 Vulpecula to 13 Vulpecula to 14 Vulpecula with four jumps of the telescope. Now let's go outside and see if this crazy stuff really works.

Okay, find Alberio. Got it in your finder? Good, now switch to the main scope and take a couple of minutes and let your retina soak up the beauty of this double star. The colors are striking aren't they? The brighter of the two stars is a deep, golden yellow and the dimmer of the two a brilliant sapphire blue. These two stars revolve around a common center and are about 410 light years from Earth. Now, go back to the finder and locate 10 Vulpecula. Is it nicely centered? Great! Take a look through the scope again. It's a beautiful field isn't it. This area is on the rim of the Milky Way at one of the great rifts caused by dark matter in the arms of the galaxy. Back to the finder again and locate 13 Vulpecula. Got it? Okay, take a look. Another nice field. 13 Vulpecula is a double star but the two stars are too close together to be separated by anything smaller than a 6-inch telescope. If you are using a 6-inch or larger scope, by all means, enjoy the view. Okay, take one last trip to the finder and move over to little 14 Vulpecula. Ready for the grand finale? Take a look. With 14 centered in the eyepiece, the Dumbbell should be at about 1 o'clock. See it? Take your time. As you move your eye around the field you will see a slight blur. That is it. Relax and enjoy. As your eye becomes more adapted you will see it better and better.

Okay, lesson over. (I'd like to see a computer do it that way!)

TELESCOPIC IMAGES

When you first look through your telescope everything is going to look wonderful. The star images will be crisp, tiny points of light, star clusters will fill the field, and the Moon will look unbelievably beautiful. But then, sooner or later, you are going to consult a star chart or a map of the Moon and the confusion will start. Nothing will match! The stars all seem to be in the wrong places. What you see on the chart is nothing like what you see through the eyepiece. Have you gone crazy?

No, not hardly. When you look at an object in the sky with your naked eye or a pair of binoculars, the orientation is correct. Look at an object through a telescope however, and everything is upside down (Fig. 6-7)! The simple explanation for this confusion is that most astronomical telescopes invert the image that you see. Because the direction of the light that makes up the image is changed in some way by the telescope's objective, either reflected by a mirror or refracted by a lens, the final image that reaches your eye is flipped end over end. Hold a magnifying glass up to view a lamp across the room and you will see what I mean. The image as it enters the lens is right side up, but as you move your eye back from the lens to the point of focus the image flips over. The bottom has become the top, the top the bottom. This flipped image is the normal way of seeing the universe through a telescope.

Problems start when we add surfaces from which our image will bounce off. When you put a star diagonal into a refractor, the image changes again. You still have the erect image but you have changed something else—the image is now reversed. Left becomes right and right becomes left. So now your image is erect but reversed. This can cause a great deal of confusion when you are trying to read a star chart at the telescope. You can always turn a chart upside down to conform with the telescope's inverted image, but there is no way you can move a chart around to make it conform to an inverted and reversed image.

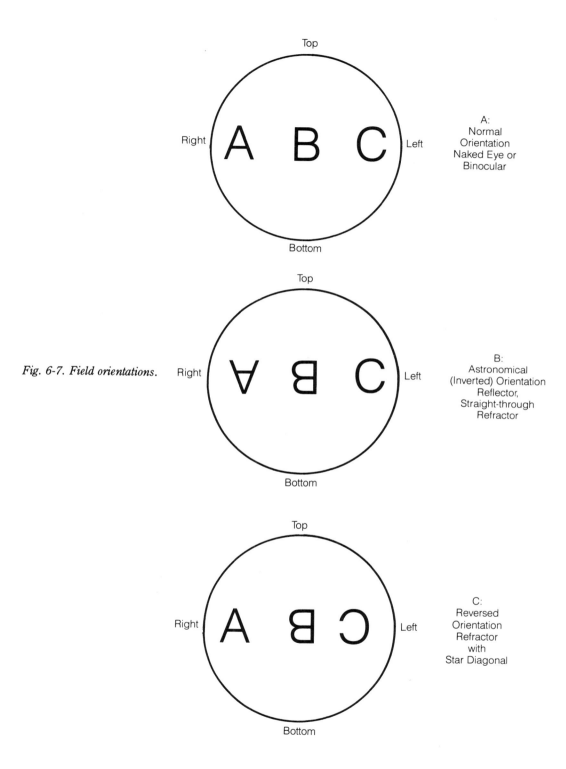

Fig. 6-7. Field orientations.

A:
Normal
Orientation
Naked Eye or
Binocular

B:
Astronomical
(Inverted) Orientation
Reflector,
Straight-through
Refractor

C:
Reversed
Orientation
Refractor
with
Star Diagonal

An image will reverse its orientation each time its direction is changed by an optical surface. The number of these directional changes is the key to what you finally see. Even numbers of changes keep the image orientation normal—left to left, right to right. A newtonian reflector gives the image two directional changes—one when it reflects off the primary mirror and another when the image reflects off the secondary—so the image is not reversed.

A refractor changes the direction of the image by bending it as it goes through each lens in the lens system. Remember that all astronomical refractors have at least two elements in their lens system. Two elements, two changes of direction, and the image is properly oriented for an astronomical instrument. Top and bottom are inverted, but left and right are okay. Add a star diagonal and you are adding a third optical surface that will change the image direction. Odd numbers of direction changes are what causes image reversal. A tri-schiefspiegler uses three mirrors to bring the image to the eyepiece, three changes of direction, making a reversed image. A Herschelian telescope reflects the image directly to the eyepiece with no secondary mirror resulting in one reflection, one direction change, and a reversed image.

You can remedy the problem of using a star chart with a reversed field instrument in a number of ways. You can turn the chart over and shine a light through the back of the chart. If you use charts that are in book form, like *Uranometria*, you will have to copy the desired chart by tracing or photocopy. This is fine, but it can be a bit cumbersome what with having to hold a chart, a flashlight, and still look through an eyepiece.

An easier solution is to find and use an *amici diagonal*. This diagonal adds an additional surface so that a refractor has an even number of image direction changes and the problem of image reversal is eliminated. There is some loss of brightness due to the extra surface, but many observers think the proper orientation is worth the price. You could take this one step further and use a system that corrects for all of the image changes by putting top on top, bottom on bottom, left to left, and right to right. But these systems, called erecting systems, waste too much of the light that enters the telescope.

Now that we have the facts straight about image orientation we're ready for our first night observing. Right? Wrong! There is one more thing we have to tackle before going to the backyard. We haven't licked this image orientation thing yet, we have to find celestial direction in the eyepiece!

Stars appear to rise in the east, transit the meridian to the south and set in the west (Fig. 6-8). This all seems cut and dried and very simple to comprehend, but remember the normal telescope field is inverted and sometimes reversed. Directions can become very confusing if you aren't prepared. The night sky is a very big place and it is easy to get lost!

Remember our inverted field? Let's give those tops and bottoms some concrete directions. The top of an astronomical telescope field is always south and the bottom is north. This may seem confusing but there are a couple of tricks you can use to get your bearings. You can find celestial north by giving the telescope a nudge toward Polaris, the North star. The direction from which stars enter the field is north. In a nonreversed field, the east is left and west is right, but in a reversed field, east is right and west is left. You can always find east by simply watching which way stars enter the field when the tele-

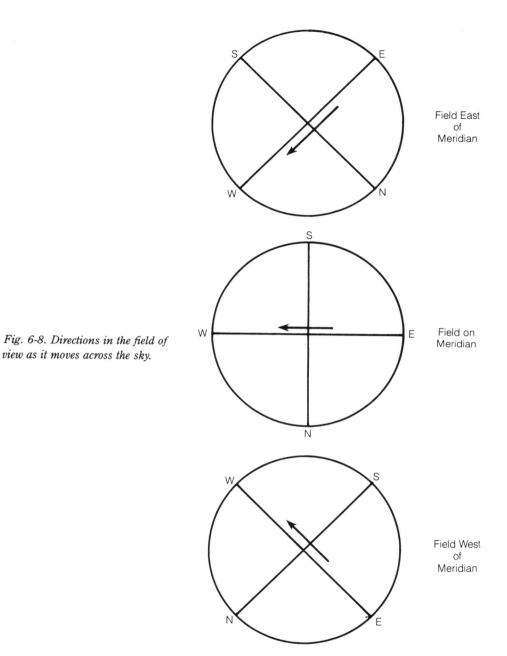

Fig. 6-8. Directions in the field of view as it moves across the sky.

Field East
of
Meridian

Field on
Meridian

Field West
of
Meridian

scope is stationary. It's not a bad idea to check your directions occasionally during an observing session. When using a chart at the telescope make sure that you orient it so the north on the chart corresponds to north in the eyepiece. If you do this everything else will fall nicely into place and you won't get lost.

✳7✳

Your First
Night Out

A journey of a thousand miles begins with a single step.

Chinese Proverb

It has finally arrived! The darkness you have been waiting for so patiently is finally descending. You have taken the time to set up your telescope in a protected area of the yard. You've spent the last hour or so checking and rechecking the telescope's collimation. You sit back on your lawn chair and watch with new wonder as the stars begin to glow from their heights. You also frown at the broadening glow of the sodium vapor lamps that suddenly seem to surround you and what you thought was a light-secure spot.

As the twilight fades you are able to pick out the outlines of constellations that only a few months earlier were unknown to you. The bright point of light that hovers over your neighbor's roof is Jupiter, it has to be! Nothing else gets that bright, that high in the sky. Almost overhead, you can discern the outline of Andromeda, the constellation that holds M31, the Great nebula! Falling toward the western horizon, but still easily visible, is Lyra with Epsilon, the famous ''double double,'' and the beautiful Ring nebula, M57! There is so much to see! Where do you start?

Start with the atmosphere. It lies between you and your goal of the stars, and is the one thing critical to both what you will observe and how you will observe it.

THE ATMOSPHERE

You wouldn't normally think of the atmosphere as a dynamic thing. It is something we all take for granted as we move through it. But take a good hard look. There seems to be something more going on than just clouds, rain, and snow. On a hot summer day take a look down your street. The air just above the street seems to be alive, shimmering and

A

B

C

Fig. 7-1. What you look through can have an effect on what you see. Warm air rising from buildings (a) can cause the telescopic image to boil. Looking across grass (b) or trees minimizes this turbulence. The best seeing is often obtained by looking across a body of water (c).

shifting like a living thing. Heat rising from rooftops in the cold of winter creates the same movement. The air is heated by either the hot street or warm roof before it rises into the atmosphere. If you try looking through that rising hot air with any size telescope, you will not see much of anything (Fig. 7-1).

The atmosphere can be your greatest help in the pursuit of astronomical wonder, but it can also be your greatest enemy. You have to know when to sit and work with your helper and when to sit back and wait for its mood to change. From the standpoint of the astronomer, the Greeks had it figured just about right. The heavens and the weather were the domain of the god Zeus, the cloud gatherer. He would let the rains fall at his pleasure. He would decide when the clouds would obscure the stars and Mount Olympus from the prying view of man. Now we have meteorologists, weather maps, and computers. They may be more accurate, but the Greek version seems more colorful.

When the clouds fill the sky it is easy to see that observing is out for the evening. It is much harder to evaluate what seems to be a perfectly clear sky. What is needed is a way to measure the quality of the atmosphere. There are, in fact, two measures of sky quality.

TRANSPARENCY

Transparency is a measure of how bright the stars appear. Another way of defining transparency is the clarity of the sky. On a good night, an observer away from the city may see stars down to magnitude 6, while another observer in the city may only see stars to magnitude 5. The difference of only one magnitude seems insignificant, but of the 6,000 or so naked-eye stars, over half of them fall below the fifth magnitude. It stands to reason that if you can see fainter stars with your unaided eye, you can see fainter stars with the help of your telescope.

I have heard stories of observers pushing their instruments past the limits set for them on paper. I have heard of observers under perfectly transparent mountain skies pushing a 60mm refractor past its limiting magnitude of 10.9 to magnitude 12, a startling feat that, unfortunately, is not likely to happen under my Chicago skies. The lack of transparency in city skies affects the limiting magnitude of telescopes as well as the eye. On a good night my 80mm refractor may pull in enough light to easily see magnitude 11 stars. While on a poor night I may find only magnitude 9 stars visible. Limiting magnitude no longer seems to be a hard and fast rule, but one that changes with the prevailing conditions.

As a measure of the clarity of the sky any haze, smoke, smog, or dust particles will work to reduce transparency. Another factor is the amount of light "splashing" up into the sky. The light hits the particles suspended in the atmosphere and lights them up, thus causing a kind of artificial aurora. This is of particular concern to observers who live in and near large cities. The southern sky from my observing location on Chicago's southwest side is often plagued by city lights bouncing off a combination of haze, smog, and dust. It has gotten so bad that I consider this to be another horizon, a "light horizon." I can see nothing below it, no matter how good the overall transparency.

Smog, dust, and haze hang in the atmosphere and there is nothing you, as an observer, can do about them. There are ways to overcome the problems presented by those pesky streetlights, however. First, if you try, you can almost always find a spot sheltered from these pesky lights. A tall tree or bush, a convenient garage, or a tall fence can serve to block the entry of extraneous light. If it's impossible to find such blocks, you can always go the route of Sir William Herschel and drape a hood over your head while at the eyepiece. I've tried it and recommend it highly. The amount of light cut down by a good, dark cloth draped over your head is phenomenal. If you don't want to go to such extremes, just cup your hand around your eye while you look through the eyepiece. This will block most of the light from the street and improve the telescopic image immensely.

There is, however, one source of light that it is impossible to overcome: the Moon. When the Moon is full, it has a magnitude of -11 and its presence in the sky washes out any faint objects such as galaxies, nebula, or faint comets. With even a first or last quarter Moon the transparency suffers to such an extent that observing deep-sky objects becomes futile. Added to the lights of a large city, the presence of the Moon will probably limit your observing to the Moon itself, planets, and bright double stars.

SEEING

The other measure of sky quality is called *seeing*. This refers to the steadiness of the atmosphere you are looking through and its resultant effects on the actual telescopic image. Seeing is affected primarily by heat and turbulence in the atmosphere. Hot air rising from a street or roof causes a telescopic image to boil, rendering it useless. The boiling causes the image to become blotted and deformed until it is impossible to distinguish anything but a featureless blob through the eyepiece. Telescopic images also tend to look jittery under conditions of poor seeing. There may be little if any deformation, but the image under observation looks like it is doing the hula! The naked-eye counterpart of the effects of poor seeing is what we all see as the twinkling of the stars.

Turbulence tends to be of two types. High altitude winds can push around layers of hot air making seeing that much harder to predict. Unfortunately there is nothing you can do about this problem. Turbulence closer to the ground, say up to 1,000 yards off the ground, can also cause these bubbles of warm air to move around and ruin a carefully planned observing session. One good thing about this lower level turbulence is that it usually dissipates by about midnight, so you can easily wait out the turbulence. Winds and turbulence at ground level can also cause the telescope to vibrate and the image to deteriorate. You can take actions to protect your telescope, and yourself from these ground level winds by finding an observing spot with a wind break of some kind.

Finally, turbulence and heated air can combine to swirl around inside the telescope tube and knock seeing all over the place. Of all the potential problems concerning seeing this is the easiest to remedy. Set your instrument outside well before you will actually begin using it. This gives the temperature of the air inside the tube time to stabilize to the outside temperature and will help improve seeing.

MEASURING SEEING AND TRANSPARENCY

E. M. Antoniadi, a Turkish-born French astronomer of the early twentieth century devoted his observational life to the study of the planets. He understood the importance of being able to know under just what conditions a given observation was made. In the later years of the nineteenth century, Antoniadi, working with the great French astrono-mer Camille Flammarion, discovered faint, elusive markings on the planet Saturn. This caused quite a stir at the time and the two were accused of fraud in some circles. It was not until the great and keen-eyed American observer E. E. Barnard saw the same mark-ings that they were vindicated.

In the 1920s, Antoniadi first proposed that the so called "canals" on Mars where only illusions caused by the eye linking details on the planet's surface to give what looked to be a meaningful pattern. In each case the observations depended on stable seeing conditions existing at the time of the observation. From his experiences Antoniadi devel-oped the following scale for seeing conditions:

I. Perfect seeing, without a quiver
II. Slight undulations, with numerous moments of calm lasting several seconds
III. Moderate seeing, with large tremors
IV. Poor seeing, with constant troublesome undulations
V. Very bad seeing

Although this scale was developed primarily for use by planetary observers it is very adaptable. The Webb Society, a worldwide group of deep-sky object observers, uses this scale to keep its records. The Association of Lunar and Planetary Observers (ALPO) uses a system that measures seeing on a scale from 0 (worst) to 10 (best). This method is a bit more subjective that Antoniadi's and makes much of the individual observer's experience.

Measuring transparency, on the other hand, is a bit easier and more standardized. Most observers rate transparency by the faintest star they can see with the naked eye. The scale runs from 0 (nothing but very bright stars like Sirius visible) to 6 (sixth-magnitude stars seen easily). Like everything else in astronomy and life, the more expe-rience you gain, the easier it is to measure transparency and seeing.

OTHER ATMOSPHERIC TRICKS

Probably the biggest trick the atmosphere plays on the observer is the simple fact that the conditions that give good transparency and seeing do not occur at the same time. For the most part a steady atmosphere with good seeing will tend to be a bit hazy. The haze acts to calm the atmosphere and improve the amount of detail an observer will see. On the flip side, a really transparent sky can set the stars twinkling like crazy with the tele-scopic images blobbing out rapidly and regularly.

The best seeing takes place just after a low pressure system when the barometer is on the rise. High winds tend to make a sky more transparent as they push warm, hazy air away from an area like a city, but then the observer has to contend with the resultant loss of definition and resolving power caused by the increase in turbulence.

You can, and should, try to fit your observing to the prevailing sky conditions. On nights with transparent but unsteady seeing, try to limit your observations to faint, extended objects that do not call for detailed examination. Galaxies, gaseous nebula, and planetary nebula fall into this category. Because they have little surface detail they are not affected by a lack of steadiness and benefit from the greater clarity brought by high transparency.

On nights where the steadiness of the atmosphere is great, study the Moon and planets with high magnification. Double stars are also good things to study during times of high stability because of the high magnification usually required to split them. If you find yourself blessed with the rare combination of high transparency and high steadiness try to contain yourself. Now is the time to do all of the above mentioned items and split some of those very close and very unequal double stars.

All things considered, the best part of the sky to observe runs from about 45 degrees above the horizon to zenith, the point directly over your head. Below 45 degrees the more atmosphere the light from the object under observation has to pass through to reach your eyes. Here light is subject to what is called *atmospheric extinction*.

Atmospheric extinction can cause problems when you are trying to find a faint deep-sky object or to estimate the magnitude of variable stars. A star loses some of its brightness when its light is absorbed by the atmosphere, making it difficult to accurately compare the magnitudes of a variable star and its comparison stars. It also makes extended objects like galaxies and diffuse planetary nebula appear dimmer than they actually are and, therefore, more difficult to locate and study. Table 7-1 indicates the magnitude loss for various angles above the horizon.

Table 7-1. Atmospheric Extinction

Degrees Above Horizon	Magnitude Loss	Degrees Above Horizon	Magnitude Loss
43	0.1	13	0.8
32	0.2	11	0.9
26	0.3	10	1.0
21	0.4	6	1.5
19	0.5	4	2.0
17	0.6	2	2.5
15	0.7	1	3.0

THE ART OF SEEING

Sherlock Holmes often admonished his friend and colleague Dr. Watson about his seeming lack of observational powers. When confronted with a seemingly simple piece of evidence Holmes would purse his lips and shake his head. "You see, but you do not observe!" To Holmes' trained eye the key details were obvious, but Watson saw nothing. Observational astronomy is very similar. The trained eye sees a wealth of detail, but most people are left to scratch their heads and wonder what they are really seeing.

The ability to see minute detail in the swirling clouds of Jupiter, to split very close double stars, or to discern faint galaxies must be learned. Sir William Herschel went so far as to warn the new observer that he must not expect to "see at first sight." To Herschel seeing was an art, an art that had to be worked at constantly to be improved. How do you go about teaching yourself to see?

Observational astronomy is a little like weightlifting. You cannot expect to bench-press 350 pounds on your first trip to the gym. You have to develop your muscles before you lift anything of consequence. You have to develop your ability to see just like you would a muscle. Working at the telescope on a regular basis is the easiest way to improve your ability to see. The more you use your ability to discern detail, the stronger that ability becomes.

The first thing you must learn is patience. You can spend a lot of time at the eyepiece waiting to see detail in Jupiter's clouds. The atmosphere is unsteady and moments of really good seeing are rare and fleeting. They are there, but you have to be patient.

Looking for detail is also a bit more involved than popping in an eyepiece and racking the image into focus. Using Jupiter as an example, begin with a low magnification. Study the surface as closely as you can without straining. As you observe, you will notice that the image may be fuzzy for 5 or 10 seconds, then become razor sharp for a second or so, then retreat back to fuzziness. That's our friend the atmosphere throwing in his two cents worth.

Try varying the magnification from time to time. Jump from your 26mm lens to your 12mm and then use the 12mm with your $2 \times$ barlow. The detail is there and it will reveal itself to you slowly.

The eye works like a camera. It tends to record the detail over a long period of time. Remember the interesting item about how the retina can selectively filter what it is viewing? This ability comes into play here. The retina initially passes along all the images it takes in until it sees those little flecks of detail that you want so bad. It then starts to filter out those fuzzy images and passes along only the sharp images that show the detail. In this respect the eye is much better than a camera. The camera only records a single moment in time while the eye records a large number of images as a single moment. You won't remember that you stared at Jupiter for 15 minutes as it floated fuzzy and blurred in the field of view. You will remember the slight wisps of white that floated on the southern equatorial belt, even though you saw it for only an instant.

Now that you are a patient observer, it's time to start talking to yourself. Get in the habit of asking yourself about the object you are viewing. If you are looking at the Moon or a planet, study it under different magnifications. Take notice of what the varying magnifications do to the image. Which makes it easier to see; which more difficult? Does a higher magnification bring out more detail or does the detail seem to get lost in the background of the scene? Once you have discovered a significant bit of detail with a high magnification, can you find it easily with a lower magnification? Ask yourself to describe, out loud, what you are viewing. For some reason this sets what you see firmer in your mind. If you are looking at a deep-sky object like an open cluster ask yourself what is its shape? Where is it brightest, dimmest? Is there any trace of nebulosity visible? What is its shape? You get the idea. By looking for specific things in the object you are observing, you will gradually see more and better detail. By talking to yourself you will become a better observer.

You can train your eye in other aspects of telescopic seeing. One important ability is to be able to judge what is referred to as *angular measurement*. This is useful in estimating distances and sizes. Whether it is an estimate of the diameter of a distant galaxy, the size of a feature near the Great Red Spot on Jupiter, or the separation of a double star, it is important to have a working grasp of the way these measurements actually look in a telescope. Angular measurements are made in arcminutes and arcseconds. Texts on astronomy define an arcminute as 1/60th of a degree and an arcsecond as 1/3600th of a degree. Fine, but what do these things look like?

The best way to develop your ability to judge angular measurement is to look at a number of objects of known size including double stars with known separations. Table 7-2 lists a number of objects and their angular size with which you can practice. It would be best to use the same eyepiece for the initial observation of each object. I would suggest starting with a 25mm. After a bit of practice, you will be able to transfer the apparently abstract concepts of arcminute and arcsecond to the universe you see through the eyepiece.

Table 7-2. Angular Size

Measured by your hand held at arm's length:

thumb to forefinger	20 degrees
forefinger to little finger	15 degrees
across your knuckles	10 degrees
your index fingernail	1 degree

Measured by sky objects:

the entire Big Dipper	25 degrees
distance between the top stars of the bowl	10 degrees
distance between the pointer stars	5 degrees
diameter of the full Moon	30 arcminutes
Mizar to Alcor (at the bend of the handle in the Big Dipper	12 arcminutes
Epsilon-1 Lyrae to Epsilon-2 Lyrae	3.5 arcminutes

As soon as you look up at the stars, you will become aware that all stars do not shine with the same intensity and brightness. As we saw in chapter 2, the brightness of the stars is set on a numerical scale ranging from the Sun at magnitude −26 to the faintest stars seen by the largest telescopes at about magnitude 25. Each step on the magnitude scale is 2.5 times brighter than the one that follows it. Accurate estimates of the brightness of an object are important when examining such things as variable stars and newly discovered comets. The best way to train the eye to estimate magnitude is, again, simply by persistent observation. In order to facilitate the development of this ability select an area of the sky for which you have a magnitude chart. Those supplied by AAVSO for variable star study are excellent for this purpose. I have reproduced one here for you to use (Fig. 7-2). These charts come with an inverted field and are ready for consultation right at the telescope. They also list the magnitudes without decimal points—magnitude 7.4 is listed as 74 and so on—so you won't confuse the point with stars. Begin with a 20mm or 25mm eyepiece and locate the area on the chart. Examine the area carefully and put a little checkmark by each star you can easily see. Notice how the brightness of the stars relate to each other. Use a slightly higher magnification eyepiece and you will see the effect increased magnification has on brightness. Can you see fainter stars? Now put the chart aside. Can you pick out the star at magnitude 6.6? How about 9.3? After a while the ability to estimate magnitudes gets to be second nature.

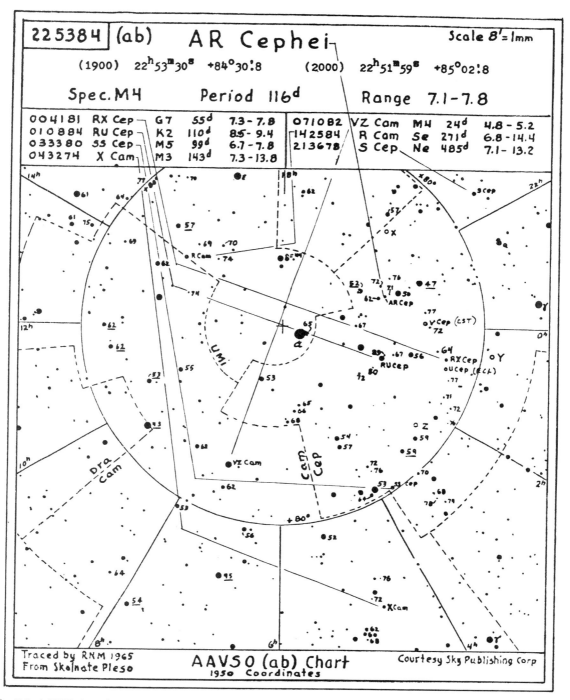

Fig. 7-2. AAVSO chart for the circumpolar AR Cephei. (Courtesy AAVSO.) (Reprinted by special permission of the American Association of Variable Star Observers, through its Director Janet Mattei.)

PROTECTING YOUR VISION

Once you have adapted your eyes to the darkness, you have to guard against any sort of bright light. If you consult a chart, use a flashlight with a red shield over the lens. This does not destroy the visual acuity you have built up because the visual purple in your eye will not react to red light. Photographers see the same phenomenon in the darkroom. Most darkroom safelights are built around a red-filtered light source. Black-and-white film and most photographic paper, like the eye, is not as sensitive to this type of light. A simple way to produce a light for chart reading is to remove the reflector from behind the flashlight bulb and then paint the bulb itself with red nail polish. The resulting light is diffused but quite adequate to check a chart or make a notation.

You also have to keep your eyes healthy and your visual acuity high. Make sure that your eyes get an adequate blood supply while observing. Alcohol and nicotine are two drugs that cut down the flow of blood to the eyes. They should be avoided especially before and during an observing session. Blood pressure is another thing that can impair vision, so keep that in check.

Other factors that enter into your own "tranquillity of the retina" include the comfort of your observing position. Sitting is the best posture, but unless you are using a refractor with a zenith prism this is difficult. Try not to tense your body while you work at the eyepiece. It is helpful to keep both eyes open while observing. This is easy if you are in a completely dark environment, but if you are observing from the city it is hard to keep distractions from gaining the sight of the unused eye. The easiest way to avoid this problem is to use an eyepatch. You can find them at most drugstores for about $1.50, a very worthwhile investment.

Next, you have to keep yourself warm. As you observe on a cool autumn evening your body heat melts away and your circulatory system has to work harder to keep your temperature stable. This is even more of a problem during the winter. The colder you are the less blood reaches the eye and your vision suffers. It is not a bad idea to invest in a good set of thermal underwear. Dress in layers of clothing as this allows layers of warm air to further insulate your body. A hat or watch cap will help retain most of the heat lost from the top of your head and add to your overall comfort. I find a thermos of tea a blessing during a session with temperatures anywhere below 40 degrees, as long as I keep it away from my charts. A thermos seems to know it has to spill on anything important. I believe it is some law of physics.

You will find that the amount of detail you are able to see decreases over the length of the observing session. This is because your eye gets tired after being used so much and tends to rebel. You may suddenly find it difficult to keep an object in focus and the strain may give you a headache. The eye ignores any image that remains in the same place on the retina for any length of time. In order to compensate for this the brain orders the eye to look harder for the image. Eyestrain is the result. In order to overcome this problem all you have to do is keep the eye moving around the field under observation. It is a good idea to give your eyes periodic rest breaks. One method is to switch eyes making sure that each gets a chance to rest in total darkness under your eyepatch.

GREAT EXPECTATIONS

When you bought the telescope at the garage sale, the guy said he was selling it because he couldn't see anything over the glare of city lights. You contend with the same lights however, and you can see plenty of stars! Could it be that there is something more than lights? If you could talk to him tonight, with the telescope all set up and ready, he would probably tell you that there must be something wrong with the telescope. It shows the Moon fine, but when you look at Mars you can't see the canals. Jupiter doesn't show the multicolored bands across its face and Saturn's rings just don't look right.

Fig. 7-3. Photos of Jupiter like this one taken by Voyager I, *have lead many people to expect more from their telescopes than they can deliver. (Courtesy NASA.)*

The person who sold you the telescope suffered from one of the most common afflictions known to the amateur astronomer. He expected too much from a small instrument. At a recent observation session held by a local astronomy club, a little girl squinted through the eyepiece of a 60mm refractor. She chewed her lower lip as she looked at Jupiter through a low-power eyepiece. Climbing down she sighed and shook her head.

"What's wrong," I asked. "Didn't you see Jupiter and its moons?"

She shrugged. " I guess so, but it doesn't look like the stuff on TV!" Our little friend is suffering from the same problem as the guy who sold you the telescope: reality not meeting expectations.

We have been spoiled by our advanced technology. The views that *Voyagers 1* and *2* sent back as they whizzed through the outer Solar System have given us a preconceived image of what Jupiter and Saturn should look like (Fig. 7-3). When we approach a telescope, be it a 60mm refractor or a 17.5-inch dobsonian, we feel that those are the images we should see. The reality of the eyepiece then cannot be anything but disappointing.

So the very first thing you must do as you approach the telescope for the first time is clear your mind of any preconceptions. If you don't, you will be disappointed regardless of what you see. Approach with an open mind and the wonders displayed before you will reward your patience and imagination tenfold.

So what do you do now? Get your telescope ready and follow along as we begin our journey through the universe.

✻8✻

Observing
The Moon

...we shall find the Moon a wonderful object of study.

Rev. T. W. Webb

The Moon is a unique object for the amateur astronomer to study. It is visible some-where in the sky almost every night. Outshining everything in the sky but the Sun, it is easy to identify. The Moon is easy to find and, above all, it is the closest celestial object to Earth with the exception of meteors. With the smallest telescope, or even a pair of binoculars, the Moon yields more detail than can be seen on any other object in the sky. With all this going for it, it is understandable that the Moon is usually the first thing toward which amateurs turn their telescopes.

For 360 years the Moon was a special place for humanity in general and amateur astronomers in particular. All that changed on July 20, 1969.

For some reason, as soon as Neil Armstrong placed his foot on the surface of Mare Tranquillitatis the Moon ceased to be an object of mystery and became just another place on humanity's vacation list. It went from being the ''Moon'' to being the moon. The missions of the *Surveyor, Lunar Orbiter,* and *Apollo* programs mapped the lunar surface so well that professional astronomers lost almost all interest in looking at the Moon with Earth-based instruments (Fig. 8-1).

This filtered down to the amateur ranks almost immediately. Prior to the age of the *Lunar Orbiter* and *Apollo, Sky & Telescope,* the premier magazine of American amateur astronomy, carried a monthly column for amateur lunar observers. Almost every month articles on specific features on the Moon could be enjoyed by enthusiastic amateurs. Then, suddenly, they stopped. Technology had triumphed again!

Selenography, the study of the Moon, has gone through phases over the 380 years since the invention of the telescope. After a period of intensive professional work, usually brought on by a technological change, interest in the Moon virtually disappears from the professional ranks. The first technological advance was, of course, the invention of the telescope itself. After the pioneering work of Galileo, the first lunar mapper Hevelius, and others of the seventeenth century, the study of the Moon fell out of favor with

Fig. 8-1. Surveyor III *on the lunar surface as photographed by the crew of* Apollo 12. *(Courtesy NASA.)*

professional astronomers. Then, with the introduction of good achromatic refractors and micrometers for detailed measurement, the study of the Moon took off again in the early 1800s. Then once again the study of the Moon fell on hard times.

In the last decade of the nineteenth century, large telescopes combined with the new science of photography to signal the start of a new age of lunar exploration and, unfortunately, an eventual decline in professional interest. Finally, the advent of space travel kindled the hopes of professionals who sought the ultimate platform from which to study the Moon: the Moon itself. That hope was realized six times between 1969 and 1973. We haven't gone back since and darkness has fallen once again on the study of the Moon.

It is during these periods of inactivity however, that the quality of work done by amateurs rivals and sometimes surpasses that done by the professionals. During these lulls the Moon is returned to those who most appreciate her: amateur astronomers.

THE DOUBLE PLANET

The Earth-Moon system is unique in the Solar System and possibly in the entire universe. From Venus or Mars, Earth would appear to be a double star, a blue-green star with a smaller white attendant circling it. It has been postulated that if the Moon circled Venus instead of Earth, we would have a totally different type of civilization. Seeing the motion of the Moon around Venus would have been a great clue to the scientists of ancient Greece that the Earth revolves around the Sun. If that theory had taken firm hold 2,000 years before it actually did, who knows what would have happened. It has also been postulated that if that scenario did take place and the Moon did orbit Venus, there would be no humanity on Earth to see it. Without the Moon there would be no great tidal movement and without that there would be no tidal pools from which life began its trek to land. Maybe it all worked out for the best?

For years and years scientists thought that we always saw the same face of the Moon because the Moon did not rotate. As with many things, these scientists were wrong. The Moon does rotate. Because the eastward rotation and the eastward movement of its orbital path around the Earth are equal, we just don't see it! That is why it looks like the same face is always turned toward Earth. But if you look closely, you may detect some rotational motion. You won't see anything over the course of a couple of hours however, because this is a very subtle motion.

If you follow the Moon carefully through more than one *lunation*, the complete cycle of its phases that lasts just over 29.5 days, you will see this effect, which is called *libration* (Fig. 8-2). Libration allows us to see around the corner of the Moon.

The Moon has a very stable rotation rate. Because of this its axis is stable with respect to the stars. On certain days during a lunation, the Moon's axis is tipped toward Earth by 6.5%. When that happens we can see 6.5% past the pole onto the other side of the Moon. These events are separated by two weeks, so we see the area beyond the Moon's north pole first and two weeks later we peek past the south pole. This is called *libration in latitude*.

The Moon does not follow a circular orbit around the Earth, but one that is slightly elliptical. Because of this the Moon revolves slightly faster than it rotates when it is closest to the Earth, or at *perigee*. When this happens we can glance around the eastern edge

Fig. 8-2. Lunar libration. (Courtesy Lick Observatory.)

of the Moon and see about 8% more of its surface. On the other side of its orbit, when it is at *apogee* or the point farthest from the earth, the reverse is true. Here the Moon revolves slower than it rotates and we get a look past its western edge, or *limb*. This effect is called *libration in longitude.*

We see more of the Moon than one would expect if only one side were turned toward us perpetually. You would expect to see 50% of the disk under those conditions. Because of libration however, we are able to see 59% of the disk. You can find out the amount of these librations by consulting one of three publications: *Astronomical Almanac*, published yearly by the U. S. Naval Observatory; the *Observer's Handbook*, published annually by the Royal Astronomical Society of Canada; *Sky & Telescope*, published monthly.

LUNAR GEOGRAPHY

When I was in high school I had to take Latin. From that long-dead language I learned that "all of Gaul is divided into three parts...." While this is not the *War Commentaries of Caesar*, this section could begin: "All the Moon is divided into two surface types." When you look at the Moon with the naked eye, these surface types are very apparent (Fig. 8-3). They form what has been called the "Man in the Moon," the "Old Woman," or even the "Rabbit and "Frog."

Fig. 8-3. The full Moon. (Courtesy Lick Observatory.)

The dark areas you see are called *maria* (Fig. 8-4), the plural of the Latin word for sea; *mare*. These are large flat plains formed by gigantic lava flows. If you look carefully, you will see that they are not black but a grayish color, something that becomes more evident with a pair of binoculars or a telescope. Contrasting with these dark areas are the lunar *highlands* (Fig. 8-5). They are mountainous regions that seem to contain countless craters. Everything else we see on the Moon is viewed against the backdrop of either a mare or a highland.

If the Moon is known for any one feature, it is known for its *craters*. Ask anyone on the street what they think of when you say Moon and the odds are they will say "craters." The term *crater* is really a generic term. There are many subdivisions that refer to crater type and size. There are *impact craters*, which are caused by the collision of projectiles with the lunar surface. Impact craters have certain unique characteristics that make them easy to identify. They are usually deep depressions that are almost perfectly circular with steep interior walls and gently sloped exterior walls. There is also usually a splash effect from the impact, called *ejecta*, surrounding the crater. Impact craters usually have a *central peak* thrown up by the volcanic reaction caused by the impact. An

Fig. 8-4. Mare Serenitatis is one of the many lunar maria. (Courtesy Lick Observatory.)

excellent example of an impact crater is Copernicus. Another interesting feature of impact craters are the *ray systems*. The most prominent system of rays belongs to the crater tycho, which stretches 1,500 miles across the lunar surface.

There are *volcanic craters* caused by—you guessed it—volcanic action. They are characterized by flat dark floors, similar to the maria. Both the interior and exterior walls are uniform, and there is no sign of ejected matter. The best example of a volcanic crater is Plato. There are also *walled plains*, *ring plains craters*, *craterlets*, and *crater pits* (in order of descending size).

In addition to craters are a number of other interesting features on the Moon. Primary among these are the *mountain ranges* that cross the surface. The finest example of a lunar mountain range is the Apennines, which stretch in an arc from just west of Plato, southeast along the rim of Mare Imbrium, to Eratosthenes. Near the Apennines is the

Fig. 8-5. The area surrounding the crater Tycho is an example of the lunar highlands. (Courtesy Lick Observatory.)

Alps, another excellent example of lunar mountains. Starting next to Plato they stretch southwest to the crater Cassini. The Alps are broken up by a representative of another lunar feature, the *valley*. The Alpine Valley runs for 110 miles through the rough terrain of the Alps. Its floor is flat and the valley itself is perfectly straight.

The Moon is also home to a number of smaller, interesting features. *Rilles* appear as cracks in the surface. Most of them are small but a few, like the Ariadaeus rill near Mare Tranquillitatis, can run for hundreds of miles. While rilles are depressions in the lunar surface, *ridges* are upswellings. They are found near the edges of the maria and suggest a wave of lava frozen in time. *Scarps* fall somewhere between ridges and mountain ranges. They are higher than the ridges, but not as high as the mountain ranges. Finally, we have small swellings in the lunar surface called *domes*.

These features of lunar geography might seen unique, but remember that every type of lunar feature has it analogy here on Earth. The Earth's features are subject to weathering while those on the Moon sit unchanged for millions and even billions of years. The meteor crater at Winslow, Arizona, is comparable to the lunar impact craters. Though it is eroded, you can plainly see the characteristic slope of the interior and exterior walls. If you want to study lunar volcanic craters you can take a trip to Hawaii. I would recommend this during January, especially if you live in the snow belt. Look at Mauna Loa and the other craters in the island chain. While there look for shield volcanoes as well. These are analogous to the lunar domes.

While these features were formed by similar events on both the Earth and the Moon, mountains, mountain ranges, valleys, rilles, ridges, and scarps all have more common origins. They were formed by events unique to Earth such as plate tectonics, and water and wind erosion. The similar features on the Moon were not caused by these events. Their formation is related to both volcanic action and the impact of large objects with the surface.

TAKE THE TERMINATOR TOUR

As the Moon moves between the Earth and Sun at the beginning of each lunation, it catches the sunlight and its surface becomes illuminated. This illumination is the cause of the phases of the Moon we see from Earth. As the Sun rises throughout the lunar day the *terminator*, or sunrise line, divides the area of the lunar surface illuminated by the Sun from the area still in darkness (Fig. 8-6).

Fig. 8-6. The three-day old Moon and Venus. (Courtesy Lee Keith.)

An excellent project for the beginning amateur is to follow the course of the terminator as it crosses the Moon. A project like this gives the beginner a chance to become more familiar with lunar features. It does not require a powerful instrument, but can be done easily with a good pair of binoculars or a small telescope. That is one of the advantages to studying the Moon. You don't need a large instrument to enjoy the surface of our neighbor in space. In fact, many observers feel that the Moon is best seen with a small instrument so that it just fills the field of view. So let's take a trip across 238,000 miles of space and see what the Moon has to offer.

The detection of the earliest crescent Moon has become a challenge to amateur and professional observers alike. The only time that observers from Earth can actually see the Moon pass from the old Moon to the new Moon is during a solar eclipse when the Moon passes across the face of the Sun. The new Moon rises about sunset and under certain circumstances a thin crescent may be visible low in the western sky after sunset. Because of the size of this lunar crescent, its proximity to the glare of the Sun, and the fact that it sets so soon after the Sun, it is rare to see a new Moon earlier than 22 hours old.

The record for visually sighting this young Moon belongs to the famous lunar mapper and director of the Athens Observatory, Julius Schmidt. On September 14, 1871, Schmidt sighted the Moon when it was only 15.4 hours old using no optical aid. The record for seeing a young Moon with optical aid belongs to an amateur from California. On March 15, 1972, Robert Moran observed the very thin crescent Moon when it was only 14.9 hours old. He used 10×50 binoculars to make his observation.

On July 14, 1988, the last opportunity to observe a young lunar crescent, Robert Stalzer, president of the Chicago Astronomical Society, journeyed to the observation deck of the Sears Tower in Chicago. From his vantage point, he had an excellent view of an unobstructed western horizon. There were clouds at sunset, but Bob began his search through a hole that pierced the cloud cover. Using 10×20 binoculars, Bob was able to detect what he estimated to be a 20-hour-old crescent. If not for the clouds, he felt sure that the Moon would have been visible to the naked eye.

The best time to make an attempt at seeing such a young Moon occurs with the following conditions. First, the Moon should be directly above the setting Sun. This occurs most often during the early spring and late winter. Second, the Moon should be at perigee because at this point in its orbit it is moving rapidly away from the Sun so the distance between the Moon and the setting Sun is greater than normal.

In order to see this youngest of lunar crescents, you need to have a number of things going for you. An unobstructed view of the horizon is essential. The target you are looking for is going to drop below that horizon in a matter of minutes as it follows the sunset. Any viewing time you can gain by not having trees or buildings in the way is to your advantage. You also need a sky with great transparency and darkness because the thin crescent is a very dim object. Finally, you need a new Moon that will be well separated from the Sun at the time you will be making your observation. You can find this out by plotting the positions for the Sun and Moon on a star chart and simply measuring the separation. You can check the *Observer's Handbook* or the *Astronomical Almanac* to see if conditions are ripe to catch a thin lunar crescent. Good hunting!

Now that you have bagged a very young Moon it's time to see the show the Moon puts on over the course of a lunation. As the new Moon races eastward away from the Sun, its features become swamped in the rising Sun along the terminator. It is along this line that you will see most features at their best. The sharp shadows thrown by the rocky lunar structures give them an almost three-dimensional effect. Whether you use a pair of binoculars or a telescope, the Moon will never let you down as a target for observation (Figs. 8-7 and 8-8).

We start our tour with the two-day-old Moon. There are a multitude of features for the observer to study. The most obvious feature is the dark oval shape that lies along the eastern limb of the Moon, Mare Crisium. Isolated from other lunar maria, Crisium is some 270 miles across from north to south and 350 miles across from east to west. It is an excellent gauge of the effects of libration because it seems to turn a fuller, rounder shape toward Earth during periods of libration in longitude.

The Moon has always been considered a dead world, basically unchanged over the billions of years that have elapsed since its formation. But something seems to take place on the lunar surface from time to time. These events are called *lunar transient phenomena* (LTP). Since 557 A.D., over 1,500 reports of LTP have been made. They fall into three distinct categories. First, there are periods when a feature or area is hidden from view. Color changes that can cover both large and small areas of the lunar surface are another type of LTP. Finally, there are glows and flashes of white or colored light. One of the regions in which these LTPs occur is Mare Crisium. Observers of the eighteenth century said that the greyish hue of the mare seemed tinted with green during the full Moon. Rev. T. W. Webb, proponent of astronomy and author of *Celestial Objects for the Common Telescope*, reports that he and other observers have seen the floor of Mare Crisium "speckled with minute dots and streaks of lights."

South of Mare Crisium, just below the lunar equator, is the crater Langrenus (Fig. 8-9). Langrenus is surrounded by crater walls that rise 16,000 feet above the lunar surface. It is 85 miles in diameter and has a central mountain rising some 6,000 feet above the crater floor. This mountain can catch the rising Sun when the floor is still deep in shadow, causing it to gleam like a diamond against a black cloth.

Farther south you will find the crater Petavius. Petavius is a huge crater that is 110 miles across. With a small telescope you can easily see the terracing of the inner walls. The floor of the crater seems to come alive with the angled rays of the Sun. There are three rille systems that criss-cross the crater's slightly humped floor. Binoculars should show one system, but the others will require a telescope. The center of the crater is home to a mountain complex, the highest of which rises over 8,000 feet above the crater floor. The floor itself is very dark and devoid of craters.

Still farther south is the crater Furnerius. This large crater is 81 miles across and over 14,000 feet deep. Furnerius belongs to the class of craters known as *mountain-walled plains* or walled plains. Craters of this class are very large but relatively shallow. Because of this it would seem this crater would be related to Petavius, but they differ in other significant ways. Petavius has a prominent group of mountains at its center; Furnerius does not. This is the principal way in which scientists define a mountain-walled plain.

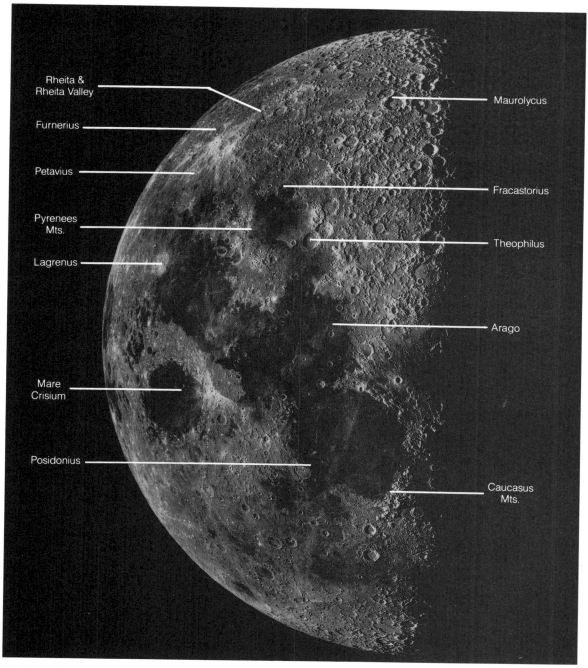

Fig. 8-7. First quarter Moon. (Courtesy Lick Observatory.)

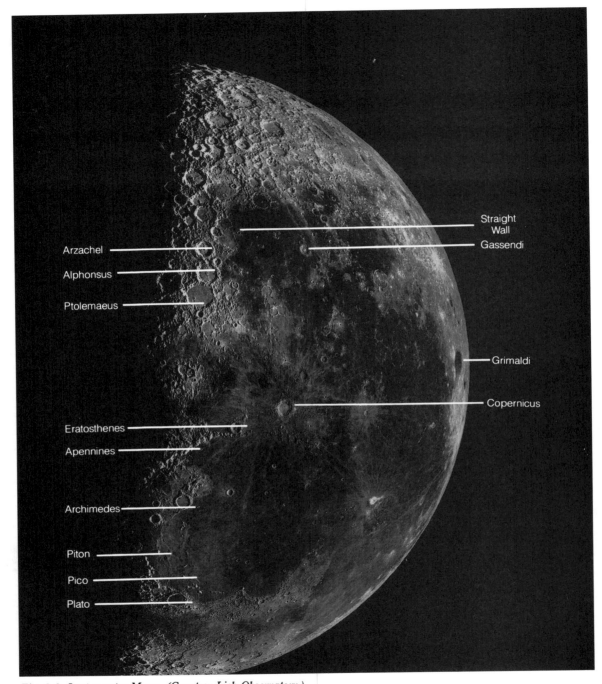

Fig. 8-8. Last quarter Moon. (Courtesy Lick Observatory.)

Another characteristic of the walled plains is their floors. The floor of Furnerius is smooth compared with the rille system that crosses Petavius' floor. Some scientists speculate that at one time the entire floor of what is now Furnerius collapsed, all at once, onto the then hot magma below it. Another piece of evidence that would seem to support this theory is the fact that Furnerius is the center of a bright ray system, as if the hot magma splashed up around the edges of the collapsed section and splattered across the landscape. Regardless of what caused the smooth floor and the ray system, it is evident that this is an old crater. Its walls are pocked with other craters, which would show that Furnerius is older than the intruding, fresher craters.

Moving inward from Furnerius you will come to the crater Rheita and the Rheita valley. Of the two features, the valley seems to be older. If you look carefully, you will see that the crater seems to overflow through the walls of the valley, an indication that the valley came first and the crater followed. The crater Rheita is smaller than the other features we have looked at up until now. It is only 44 miles across and 14,500 feet deep. The valley runs for about 300 miles and is about 2,000 feet deep.

Fig. 8-9. The crater Lagrenus. (Courtesy Lick Observatory.)

As the terminator moves west across the Moon, you will notice that the other maria, which are becoming visible, are connected. Each seems to flow into the next. One reason the old mappers thought them to be seas was because of this flowing motion, thus their name. Northwest from Rheita and the Rheita valley is a dark mare called Nectaris, or the Sea of Nectar. At the south end of this mare is our next stop, the crater Fracastorius. This crater is 73 miles across, but you will have to look closely to find it because it has lost most of its north wall. The remaining walls rise only 5,800 feet above the surface. Because of this relatively small height, Fracastorius doesn't throw too much shadow. Fracastorius looks more like a bay than a crater and the crater looks old, which it is.

Two other features frame Mare Nectaris for us on our guided tour across the Moon. The first of these is found along the eastern edge of the mare. The Pyrenees mountains fit better into the category of scarp than mountains. Experts from the U.S. Air Force and Army mapping services cannot agree on their height. The USAF lists the range as having a maximum height of only 6,800 feet, while the Army puts the height closer to 10,000 feet. They rise above the shore of the mare and run for about 200 miles along its rim. Whether you want to classify this feature as a true mountain range or a scarp is up to you. Whatever you end up calling it, the Pyrenees are easy to identify and a striking object in the rising shadows thrown by the Sun.

Across the mare in its northwest corner, we come to our other framing feature, the crater Theophilus (Fig. 8-10). At 65 miles in diameter and over 22,000 feet deep, Theophilus makes an impressive sentry guarding the connection between mares Nectaris and Tranquillitatis. Theophilus is the deepest of all lunar craters and has a central mountain peak reaching 7,500 feet above the crater floor. When near the terminator, the floor of the crater is lost in darkness while the tip of its mountain shines brightly.

Moving north and west from Theophilus we come to a small, seemingly insignificant crater called Arago (Fig. 8-11) sitting in Mare Tranquillitatis. The crater is the smallest we will see in this short trip across the Moon. It is only 16 miles across and 5,900 feet deep, but we didn't come here just to see the crater. Arago sits on the relatively flat surface of Mare Tranquillitatis, made famous by Neil Armstrong and company. If you look closely at the surrounding mare plain with a telescope, you may see little bumps every now and then. These are lunar domes and this area is rich with them. Directly north of Arago, about the distance of the crater's diameter, you may see a dome, called Arago Alpha. There is another dome directly west of the crater at about the same distance, called Arago Beta. Lunar mappers have taken to labeling any surface bump with the name of its associated crater and a Greek letter. Just southeast of Arago sits the old, ruined crater Lamont. Between Lamont and Arago are six objects suspected of being domes. Take a close look at these objects. Do they look like Alpha and Beta?

North of Arago on the shore of Mare Serenitatis is our next stop, the crater Posidonius (Fig. 8-12). At 62 miles in diameter it has all the characteristics of walled plains like Furnerius, but with a distinct difference. Furnerius and Posidonius both have the flat floor that is one of the trademarks of the walled plains, but through a telescope you can see that Posidonius is a jumble of activity. You can see that this crater floor is filled with numerous smaller craters, an indication of great age. The floor of the crater also contains four rille systems, one of which, Rima Posidonius II, is easy to spot with a small telescope.

Fig. 8-10. The crater Throphilus. (Courtesy Lick Observatory.)

Moving west around Mare Serenitatis you come to your first real lunar mountain range: the Caucasus mountains. These are immediately distinguishable from the Pyrenees because they are almost twice as high. It is no wonder that some observers consider the Pyrenees to be nothing more than a glorified ridge! Before the terminator even reaches the Caucasus range you can see it. The rising Sun catches its peaks and these shine against the still-shadowed Moon. This range runs for over 300 miles separating Mare Serenitatis from Mare Imbrium.

The first half of the tour is coming to an end. The Moon is approaching first quarter phase and by the time this half is over we will have given 13 lunar features a detailed look. The last stop on this part of the tour takes us back toward the south end of the Moon. Maurolycus is 73 miles across and almost 17,000 feet deep. Its size and depth seem to dominate this area of the lunar surface. On the terminator it stands in bold relief. It is a relatively young crater plainly formed by impact. Observers of the nineteenth century reported that the floor of the crater is criss-crossed by up to a dozen bright lines when the area is seen under high illumination. The central mountains of the crater are easily seen with binoculars, especially when it is near the terminator.

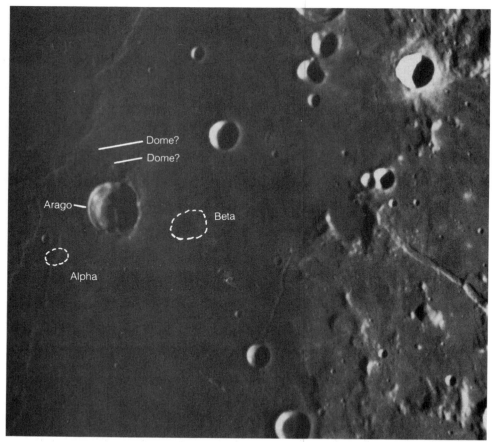

Fig. 8-11. The crater Arago and lunar domes. (Courtesy Lick Observatory.)

By now you have watched the terminator creep across the Moon at a rate of about 12 degrees per day for about five days. It is seven days past new Moon and the Moon has traveled through one quarter of this lunation, but there is much more to see.

We will begin this half of our journey just south of the lunar equator at a group of three craters. The largest of the group is the crater Ptolemaeus with a diameter of 93 miles and a depth of 9,800 feet. Looking at the crater's interior with a pair of binoculars shows that the crater floor looks smooth. Just north and east of the center of the crater however, you will notice a small, bright spot that is the small crater Ptolemaeus A. Only 6 miles across, it is the smallest crater on the tour. Through the telescope the floor of Ptolemaeus reveals a more complex structure. Craters and craterlets are everywhere. The oblique light of the terminator reveals a wealth of ridges and depressions on the crater floor.

Fig. 8-12. The crater Posidonius. (Courtesy Lick Observatory.)

At the other end of the group is the smallest of these three craters, Arzachel. It is the youngest of the group and shows all the characteristics of a classic impact crater. It is 59 miles across and 13,000 feet deep, and has a central peak rising 4,600 feet. A look at the crater with a telescope reveals beautifully terraced walls over 10 miles thick.

The crater in the middle of the group, however, is the most interesting. Alphonsus is an old crater that has been partially flooded with lava, possibly thrown up by the impact of the object that created Arzachel. It is 73 miles across and only 6,600 feet deep with a central peak only 3,800 feet high. But it is not its physical characteristics that make this an interesting object.

In 1956 the lunar expert Dinsmore Alter used the 60-inch reflector at Mount Wilson to make a photographic study of the Alphonsus region. The photo of the area east of the central mountain was puzzling because it showed the floor of the crater covered by a patch of haze. A Russian astronomer, Nikolai Kozyrev, became interested in the problem of lunar changes and began observing the region with the Crimean Observatory's 50-inch reflector in November 1958. Some three months later in *Sky & Telescope*, Kozyrev revealed that he did see a change take place and that he had definite proof in the form of spectographs.

While working at the telescope and taking photos of the spectrum of light reflected back to Earth from the area of the central mountain Kozyrev saw what looked to be a gas cloud form near the central peak. It was accompanied by a reddish glow. He observed a second event during which the peak of the mountain became very bright and white. A later analysis of the spectographs show the presence of a gas cloud made up of carbon molecules.

During the Ranger photographic program of the mid-1960s, Alphonsus became the target of the last Ranger probe, *Ranger 9*, which impacted with the crater on March 24, 1965. Photos returned by the probe (Fig. 8-13) show certain parts of the floor covered by a dark material that could be ashes and cinders thrown up by a lunar volcano. Could something like this have caused the events that Kozyrev observed?

The next feature we encounter is just southwest of Arzachel. It is easy to spot this feature. Many people, when they first see it, are struck by the fact that it looks artificial. The Straight Wall is just what its name says it is: a straight wall of rock that runs for 70 miles, rising from its north end to a height of 1,200 feet, maintaining that height for 20 miles, and then dropping back to the plain. The sharp figure cut by the Straight Wall was caused when the plain on the western side of the feature slid away because of some tectonic or volcanic action. The Straight Wall is thus the demarcation line between the two sections of the plain.

Because the western side of the Straight Wall is lower than the eastern side, the feature presents a changing image to us during the long lunar day. During the period of morning, or rising Sun, the Straight Wall will appear as a dark streak on the lunar surface because the shadows are thrown from the ridge of the wall onto the lower western plain. During the period of the setting, or evening Sun, the light illuminates the exposed face of the wall and it appears as a bright streak.

If you extend the line of the Straight Wall to the northwest you will come to our next stop and what many think is the most spectacular crater on the Moon, Copernicus (Fig. 8-14). Copernicus stands out even when the Moon is viewed with the naked eye. With a little optical aid from binoculars, the crater begins to dominate the lunar landscape. With a telescope it is simply magnificent!

Copernicus is, by all measures, a rather mediocre crater, with a diameter of 60 miles and a depth of 12,600 feet. You have already seen craters bigger and deeper. So what makes Copernicus so special? Copernicus is a young crater. Compared to a structure like Fracastorius it is the new kid on the block. The relative youth of Copernicus gives it a youthful appearance. Its lines are strong, sharp, and clean, not like the collapsed wall on Fracastorius. As an excellent example of an impact crater, Copernicus has a fine ray system that was thrown up when its meteoric creator collided with the Moon. Combined with its high surface brightness and its position on a dark plain, the ray system seems to draw attention to the crater. Finally, Copernicus obeys the three basic rules of real estate: location, location, location. Take a look at how Copernicus is placed on the lunar disk. Not very far from the center, is it? Even if you aren't looking for it, your eye is going to automatically gravitate to that bright patch near the center of the Moon. It's almost a reflex action.

Fig. 8-13. Ranger IX *photograph of the interior of the crater Alphonsus just prior to impact. (Courtesy NASA.)*

Close by Copernicus, just to the northeast, is the crater Eratosthenes. Though smaller at only 37 miles across and 12,000 feet deep, Eratosthenes is similar to Copernicus in many ways. Because the Sun strikes Eratosthenes before Copernicus, it is sometimes confused for the larger crater. The area between the two craters is interesting to study with a telescope. Because both of these features are impact craters, an awful lot of material had to be thrown up during impact. The dozens of tiny craterlets that fill the plain between the two craters tell the story of where that material came down. Eratosthenes also has an interesting feature to its southwest. A small ridge runs for about 50 miles and looks like a small tail.

Fig. 8-14. The craters Copernicus and Eratosthenes. (Courtesy Lick Observatory.)

Just east of Eratosthenes and stretching northeast along the rim of Mare Imbrium is the finest example of a mountain range on the Moon. The Apennines (Fig. 8-15) extend for 450 miles from Eratosthenes to the shore of Mare Serenitatis, and rise to a height of 20,000 feet. The peaks catch the Sun well before the arrival of the terminator and hold those rays after the terminator has carried sunset away from the region. The mountains are easily seen under the full glare of the Sun during full Moon.

The eastern approaches to the great flat, dark plain called Mare Imbrium are guarded by two great sentinels. To the south is the large, dark-floored crater Archimedes. Another walled plain, Archimedes is 51 miles across and 6,800 feet deep. Its walls are low at the north and south points of the ring, almost as if some projectile went through them on a straight line. The floor of the crater is dark and the selenographers William Beer and John Madler called it "smooth as glass" in their epic 1837 work *Der Mond*.

At the north end of the approach to Mare Imbrium lies another great walled plain, Plato (Fig. 8-16). Although many observers consider Copernicus to be the premier lunar crater, Plato is my favorite. Elliptical in shape, Plato is 64 by 67 miles across and 8,000 feet deep. The floor is very dark in contrast to the floor of Archimedes. Most of these old walled plains tend to fade out in the glare of the Sun as the Moon becomes full, but the floor of Plato seems to get darker as the Sun climbs in the sky. Consequently it is as easy to see at lunar noon as it is at sunrise over Imbrium.

Although the wall structure that rings Plato is old and not as high a ring as that of Copernicus, it combines with the dark crater floor to put on a very enjoyable show during sunrise. As the Sun climbs over the wall, it casts wonderful shadows across the floor. There are sharp points and valleys that never cease to amaze the viewer. Plato has also been the sight of numerous LTPs, ranging from flashes and lights to obscuring fogs, so keep an eye out.

Between Plato and Archimedes lie two of the Moon's rare single mountains. Piton is directly south of Plato and Pico is a bit east of a line joining the two craters. Piton covers an area of almost 190 square feet and rises 8,200 feet above the floor of Mare Imbrium. Pico is a more complex structure over 18 miles across at its base and jutting upward approximately 8,000 feet above the mare.

Back across Mare Imbrium and southwest from Copernicus you will find Gassendi (Fig. 8-17). Seventy miles across and 6,600 feet deep, Gassendi offers the observer with a telescope a lot of detail. Lodged in its northern wall is a much smaller crater, Gassendi A. The floor of the crater is covered with rilles and a central peak that rises 3,600 feet.

Finally, the terminator reaches the crater Grimaldi just before the Moon becomes full. Grimaldi is a very large crater over 140 miles across and more than 10,000 feet deep. Remember how dark Plato's floor was? Well Grimaldi's is even darker. It only reflects back 6% of the light that reaches it.

The Moon has reached the phase we call *full* and for a brief period there is no terminator. But before you rush off, you have to realize that you are only half done! We saw sunrise across the face of the Moon. The features we saw at the start of the tour—Langrenus, the Rheita valley, and Theophilus—were touched by the rays of the Sun slanting in from the east. In effect you have only seen half the show! Now it's time to

Fig. 8-15. The Apennines and the crater Archimedes. (Courtesy Lick Observatory.)

watch as the Sun drops toward the western limb of the Moon and the shadows stretch out from that side of the craters. It's going to be a different show.

In one sense, the sunset tour is going to be easier. You already know the craters and the other features to watch, but in another sense it's going to be harder. Because of the way the Moon orbits the Earth, it rises about 45 minutes later each night during a lunation. When you began the tour the Moon was up just after sunset, but it has been falling behind in the sky later each night. Now that the Moon is past full it will be rising later each evening, so it's going to take a little discipline to continue the tour. By the time this lunation is over, you will be observing in the wee hours of the morning before sunrise. But trust me, it's going to be worth it! The shadows you saw thrown across the floor of Plato were cast by the east wall of the crater. Aren't you just a tiny bit curious about what kind of shadows its west wall throws?

Fig. 8-16. The crater Plato. (Courtesy Lick Observatory.)

Fig. 8-17. The crater Gassendi. (Courtesy Lick Observatory.)

THE MOON AND THE SERIOUS OBSERVER

So now you have a feel for some lunar geography. You took the entire terminator tour and can find your way around the lunar landscape pretty well. So what's next you ask?

Map and Filters

If you decide that you want to do some serious lunar observing, there are a couple of things that you will need. The first requirement is a good lunar map. You have an idea of what the various lunar features look like, but you need a means of finding your way through the hundreds of thousands of craters that dot the Moon.

Sky Publishing Corporation distributes an excellent resource entitled *Lunar Quadrant Maps*, prepared by scientists from the University of Arizona's Lunar and Planetary Laboratory. Printed in four sections, the maps show all the named features on the visible face of the Moon down to a size of two miles. Each chart is large, measuring 23 by 27 inches. The set is rolled and mailed in a heavy tube so it is a good idea to mount them on some sort of board for use at the telescope. One amateur selenographer I know divided each chart into six sections, photocopied each of those sections, put the copies into transparent sleeves, and then put them all into a ring binder. He has the original charts mounted on the walls of his family room and takes only the chartbook out to the telescope with him.

You might have noticed that looking at the Moon, especially when it is near full, is very tiring on the eyes. This is because of the tremendous glare reflected from the lunar surface. The easiest way to overcome this is to use a filter. A number of companies manufacture a special filter for observing the Moon. They are threaded at both ends so they easily screw onto the end of your eyepiece. The best filter for reducing glare is called a *neutral density filter*. Neutral density filters are gray and have no effect on the image other than to reduce the overall glare. Also available are *variable density polarizing filters* that change their darkness to admit more or less light to suit the observer.

While neutral density filters or polarizing filters may look dark, they are not designed to look at the Sun. Do not use neutral density filters or polarizing filters to view the Sun!!!

A good source for these filters is a photo shop. Eastman Kodak and many small companies manufacture a line of filters for photographic use. They are much bigger than the filters made specifically for an eyepiece but that's okay; it is just as easy to hold the filter between the eyepiece and your eye.

Wratten filters are simply sheets of filter material that you can cut to a convenient size and place between your eye and the telescope's eyepiece. If you are studying LTPs, you can use a red (Wratten 25) and a blue (Wratten 38A) to bring out the contrast in the suspected LTP. Wratten filters are expensive because they are usually sold in large, 3-inch-square sheets. Contact a local photographer and ask if he has any in his dust bin. He probably does and would be happy to get rid of the torn and broken things. Take the filters or the pieces home and mount a red and blue piece side by side in a 2×2 slide mount. You can place the slide mount up to your eye and move it back and forth, which will cause any suspected LTP to flicker.

Studying Lunar Domes

Okay, now that you have a good lunar map and a filter to cut the lunar glare, what can you do? A project you can start even before you have a telescope is a study of lunar domes. Lunar domes are those swellings we saw around the crater Arago in Mare Tranquillitatis. They are important to lunar science because they may represent a late stage of volcanic vents that could have contributed to the formation and later changes of the lunar maria. One of the by-products of the technology that took man to the Moon in 1969 was the complete photographic portrait made by the satellite that preceded the landing. The Lunar Orbiter project of the late 1960s offers a wealth of detail that anyone can study.

The *ALPO Lunar Dome Recorder* can supply you with a listing of suspected lunar dome sights plotted on the *Lunar Quadrant Map* discussed earlier. An observer can then study such atlases as *The Moon as Viewed by Lunar Orbiter* (NASA Publication SP-

20) and the *Atlas and Gazetteer of the Nearside of the Moon* (NASA Publication SP-206). Although these works are out of print, they can be obtained at most libraries. I also use a number of atlases created from Earth-based photographs of the Moon, but these are not necessary. Using material like this is easy, you don't have to worry about being clouded out!

The procedure for the project is simple and can be rather enjoyable. For equipment you will need a magnifying glass and a ruler equivalent to the scale of the photos you will be studying. Begin by locating the area on the ALPO publication that you want to chart. Each suspected dome is listed with a number so you don't have to worry about figuring out any sort of coordinates for any objects you find. Flip to the section of the atlas that has the area you wish to search and carefully go over that area with your magnifier. Take your landmarks from the ALPO chart and search away! If you find an object that corresponds to a suspected dome on the ALPO chart, note its number and look for the following details.

* Is the suspected dome in an area of lunar highlands, maria, or somewhere in between?
* What is the object's approximate size? Use your ruler for this. Is it less than 5 kilometers at its longest? From 5 to 20 kms? 20 to 35 kms? Or larger?
* What is the object's shape? Circular? Elliptical? Polygonal? Irregular? Or is it too ill-defined to classify?
* Does the object have a gentle or moderately sharp slope?
* What would the object look like if you were to view it in cross section? It is hemispherical? Would the summit be flat or sharp, or would there be multiple summits? Is the suspected cross section regular or off center?
* Are any details visible on the summit? Is it a craterlet or pit? A ridge or a peak? A cleft or valley? Or even nothing at all. How is any detail positioned on the summit? Centered? Off center? On the edge? Does the detail cross the entire summit?

Later, you can take your notes and charts out to your telescope and see if you can find the objects you found in the photos on the Moon itself.

Details on the *ALPO Lunar Dome Survey* can be obtained from the *ALPO Lunar Dome Survey Recorder* listed in the appendix.

Making a Detailed Study

Once you obtain your telescope and become familiar with it, an enjoyable and worthwhile project for the beginning observer is a detailed survey of a specific area of the Moon over the course of an entire lunation. A project like this will give you valuable experience in planning and executing a scientific study that doesn't have to rely on expensive equipment. It will also help to sharpen your observational skills, something that every amateur astronomer can afford to do.

The area that you select for your study is up to you. Try not to select an area too packed with details or you might have trouble keeping everything straight. For our example here, I am going to use my favorite area of the Moon, the area of Mare Imbrium around Plato. Let's begin by selecting the exact area in which we are going to conduct our study by examining a high-resolution photograph of the region. One nice thing about being an amateur astronomer is that there is no pressure in anything you do. You can make this as ambitious or as simple as you want; it depends on you and no one else. I want to study the crater itself over a long period of time, so I'm going to stay fairly close to the crater.

If you are going to study the crater over a period of a complete lunation, you must have some way of recording the results of each evening's observations. There are a number of ways to record your observations. Every amateur keeps a notebook or a journal with his or her astronomical observations. Some use preprinted forms, some use a simple 4×6 card for each observation, and some use a loose-leaf notebook. Some of the high-technology types have their records on computer floppy disks! However you do it, it is very important to keep track of what you see and what you do in observational astronomy. Through your records you will be able to trace your development as an observer and study both your weak and strong points. The simplest way to record an observation, be it of an area on the Moon or a distant galaxy, is by putting pencil to paper and writing. Notebooks everywhere are filled with notational sketches such as:

February 12: 02:45 UT (8:45 P.M. CST) Having a bit of trouble locating R Cnc tonight, my diagonal always confuses me! Got to remember which way is celestial north!

or

June 26: 04:00 UT (9:00 P.M. CDT) Plato looks great tonight, the floor is very dark. Piton's seems unusually bright tonight.

Many times these verbal descriptions are supplemented by simple line drawings made at the telescope. These drawings don't have to be the work of a Rembrandt. All the amateur wants them to do is convey an image of what was seen.

The best way to record lunar detail takes one of two forms: a photograph or a drawing. At first glance it would seem that a photo of a feature is the way to go, but many observers don't feel that way. The human eye may lose out in a contest for seeing a faint point of light, but it is infinitely better at filtering fine details in a scene. A photo of a lunar crater is a single instant in time, while a drawing is a compilation of many fine details seen by the eye and transmitted from the brain to the paper.

"But I can't draw. I have trouble sharpening a pencil," you may say. Well, believe me, you're not alone. The easiest way to draw lunar features is to cheat a little bit and start with an outline chart of the feature you are going to observe (Fig. 8-18). Find a good-quality photo of the target area such as the one in the NASA atlases or a photo taken by an observatory. Decide which area of the photo you are going to observe and carefully trace the basic outlines of the features you want to study. At the telescope use pencils of varying sharpness to record the fine details you see.

Don't feel like you are cheating. This technique was pioneered in the 1890s by a German selenographer named J. N. Krieger. He used this technique to compile one of the most accurate, and beautiful atlases of the Moon ever published. *Joh. Nep. Kriegers Mond-Atlas—Neue Folge.* Published in Vienna in 1912, the *Mond-Atlas* contains 58 plates of various sections of the Moon. The atlas is so rich in detail it rivals every photographic atlas I have ever seen.

Fig. 8-18. Outline of the crater Plato for lunar drawing.

Once you have your basic outline, make a dozen or more photocopies of it—one for each night that the feature is visible. Each time you observe your target, take along a blank outline and fill in the fine details as you observe. Somewhere on the form you should note the date and time, preferably in Universal Time, of the observation. Also record information about the telescope, eyepiece, and magnification used. Finally estimate the seeing and transparency for the night. Use an area somewhere on the face of the outline chart to record notes about what you are seeing to supplement the information in the drawing.

Now that you have your outline chart, pencils, and telescope you're ready to go. But what are we looking for? You can consider the basis of this exercise as being twofold. First and foremost you simply want to take some time and record what you see at the telescope. You want to become familiar with your target crater. Once you have found your target take a few minutes and let your eyes roam around a bit. Remember what I said about talking to yourself? Well it's time for that again. How does the crater look in overall appearance? Can you make out anything on the crater floor? What kind of detail can you see in the walls? How about the surrounding rough terrain? Any small cracks or clefts? Don't just talk to yourself, make notes. Get in the habit of recording everything you see. Some observers take a small recorder out to the telescope with them and use it to record descriptions. I've done it and can safely say that it is convenient and easy to use.

Another thing you can do while at the telescope is to monitor your target regions for LTP. These can occur anywhere on the lunar surface, but many of the 1,500 reported events seem to occur repeatedly in particular areas. ALPO recommends an extended study of eight areas for possible LTPs. They include Alphonsus, Aristarchus, Atlas, Copernicus, Herodotus, Tycho, Theophilus, and Plato. You will notice that our target for this exercise, Plato, is on ALPO's list so it's not a bad idea to start including searches for these events in your observation of the area. I have included lunar photo charts of the other areas in this chapter.

Studying Occultations

Another area in which the amateur can make a significant contribution is in the study of *occultations*. An occultation occurs when the Moon moves between a star and an

Fig. 8-19. Moon moving through the Pleiades during an occultation. (Courtesy Walter Piorkowski, ©1988.)

observer on Earth (Fig. 8-19). Since the invention of the telescope astronomers have used occultations to accurately determine the position of the Moon in its orbit, in respect to the time and location of the observer on Earth. Occultations also can reveal faint double stars and even help to determine the diameters of stars. Not a bad set of priorities for an observer with a small telescope!

For every occultation there are two possible types of events. Depending on where the observer is located he can see either a total or a graze occultation. A *total occultation* will occur if the observer sees the star disappear completely behind the bulk of the Moon. If the occultation is scheduled to take place along the northern or southern limb of the Moon, a *graze occultation* takes place. If the observer is positioned to watch the occultation along the Moon's limb he or she may see the star disappear and reappear among the mountains and valleys along the lunar limb. The role of the observer during total occultation is simply to record the time of the star's entry and exit behind the lunar disk. During a graze the observer should make a series of timings, one for each entry and exit the star makes as it moves along the uneven lunar limb.

Equipment needed for this work is simple. A telescope with a good mount is, as with any other type of astronomical observation, mandatory. It would be best to use an equatorial mount because the motion of this type of mount allows the observer to follow the target star easier. A clock drive is not required, but it allows the observer to focus all of his attention on timing the event and not worrying about the star and the Moon drifting out of the field.

The Moon and the Serious Observer 139

The next piece of equipment you will need is a radio that can be set to receive accurate time signals. The National Bureau of Standards maintains a 24-hour radio station that does nothing but broadcast time signals. The call letters for this station are WWV and it can be received on the shortwave bands of 5, 10, 15, and 20 MHz. If you use a tape recorder, you can make your comments and observations freely while the time signals from WWV are recorded in the background.

Finally, a stopwatch is needed to serve as a timing instrument and a backup. That's all the equipment it takes to get into occultation timings. The procedure for observing and timing a total occultation is just as simple. With the radio set to WWV and the tape recorder running, simply give a mark that can be heard over the time signals when the star either disappears or reappears.

It is easier for the beginning observer to watch and time total occultations. It is recommended that you try a few of these before you jump into a graze project. Predictions for these events for stars brighter than magnitude 5 appear in the January issue of *Sky & Telescope* and the *Observer's Handbook* published yearly by the Royal Canadian Astronomical Society (RCAS). The detailed prediction tables listed in the RCAS publication can tell you if the event is a disappearance or a reappearance against the bright or dark limb of the Moon.

The easiest occultations to observe are disappearances that take place against the dark limb of the Moon. The dark limb of the Moon allows you to find the star well before its time of disappearance. An event that is to take place against the bright limb of the Moon is more difficult to observe because the brightness of the lunar limb can make it almost impossible to locate the target star even if the observer begins to look for it well before the event.

Predictions for graze occultations can be obtained from the International Occultation Timing Association (IOTA) or the U.S. Naval Observatory. Observing and timing a graze is a bit more complicated and something best done by a group of people. A graze can only be seen from a narrow path that may be less than a mile wide. Because of this it is necessary to place observers all along a predetermined length of the path. Many astronomy groups regularly observe these events, and it would be best to take your first plunge into graze occultations with a few lifeguards present.

✳9✳

Observing
The Planets

*The planets are seen very rotund, like little full moons and
of a roundness bounded and without rays.*

Galileo

In July 1976, *Viking 1* transmitted the first live television pictures to Earth from the surface of Mars. *Viking 2* followed two months later. Astronomers, both amateur and professional, were told the days of Earth-based telescopic studies of Mars were over. Both Vikings have been turned off for almost a decade.

In early March 1979, *Voyager I* reached the vicinity of Jupiter and returned fantastic pictures of Jupiter and its moons. In July 1979, *Voyager II* repeated the triumph of its twin spacecraft. *Voyagers I* and *II* then hurried toward Saturn, arriving there in November 1980 and August 1981, respectively. Again, both amateur and professional astronomers were told the days of Earth-based studies of Jupiter and Saturn with telescopes were over. Today, both Voyagers are on their way out of the Solar System.

During the late nineteenth century, astronomy concentrated on the study of the planets. The moons of Mars and a new moon orbiting Jupiter were discovered by visual observation. Theories about the possibility of life on Mars abounded and discoveries of comets and asteroids were an almost daily occurrence. The study of galaxies lay beyond the grasp of the instruments of the time and theories of cosmology lacked observational data that made sense.

The development of the large reflectors and instruments like the spectrograph and photometer during the beginning of this century changed the stream of astronomical thought. Interest shifted to objects millions of light-years away from the Solar System. Theorizing about the formation of the universe took the place of wondering about the planets.

With the advent of the space age and the flights of robot probes to the residents of the Solar System, it would seem that all the questions have been answered and nothing remains to investigate. But nothing could be further from the truth. In astronomy, like any other branch of science, as old questions are answered new ones are raised. What is the reason for the massive dust storms that blanket Mars from time to time? What mechanism forms storms in the atmosphere of Jupiter? How are the orbits of Saturn's inner satellites and its ring system related? These are just some of the questions raised by the flights of Viking and Voyager. They were there to ask the questions and supply some of the information necessary for the answers, but Viking and Voyager were there for only a brief time.

Where will the data to answer these questions come from? "Time on our large, complex telescopes is too valuable to waste on such trivial objects as the planets," says the professional astronomer. He's concerned with analyzing the light from a distant quasar. "I guess we'll just have to wait until we can get another probe out there."

"Yeah, but our budget has been cut to pieces," replies the NASA scientist. He's concerned with another upcoming round of budget cuts. "We may not get a launch for a planetary probe for another decade or more."

"Excuse me," interrupts the amateur astronomer. "Would data on the last ten Martian dust storms be more useful to you if I put it on a floppy disk?" He's concerned about getting home in time to observe a new dust storm on Mars he heard about from his friend, an observer from Japan.

Where do you think the information will come from?

THE AMATEUR'S SOLAR SYSTEM

Before the invention of the telescope, the Solar System was comprised of the Earth, the Sun, the Moon, and five planets. By the time the telescope had been on the scene for some 240 years, astronomers had added two planets, some planetary satellites, an occasional comet, and some asteroids to the Sun's family. In 1930, Clyde Tombaugh added the planet Pluto to the stable. Ask any school child today what the Solar System consists of and you will hear the litany of the planets with a couple of asteroids and Halley's comet thrown in for good measure.

I once heard a lecture by a prominent astrophysicist about the Solar System. "The Solar System," he began, "consists of the Sun, Jupiter, Saturn, and assorted rubble." From the standpoint of the astrophysicist this is quite true because 99% of the mass of the Solar System is concentrated in those three bodies. The other planets—including Earth, the satellites of the planets, the asteroids, the comets, and everything else—comprise less than 1% of the stuff that makes up the Solar System.

Kind of makes you think, doesn't it? Each period of history and each scientific specialty has its own view of what comprises the Solar System. Interestingly enough, each view is correct.

The Solar System might consist of the nine major planets, but let's look at them from an amateur astronomer's observational point of view.

Closest to the Sun, Mercury is a small world, only about 3,000 miles across. Called an *inferior planet* (Fig. 9-1) because its orbit is inside that of the Earth's, it shows phases, just like the Moon, when viewed through the telescope. It is also because of its orbit that Mercury is one of the most difficult of the planets to observe. When Mercury is east of the Sun, we see it in the evening sky after sunset. When it is west of the Sun, we see it in the morning sky just before sunrise. But because Mercury travels in such a tight orbit around the Sun, it never rises very high above the horizon and doesn't get much brighter than magnitude −1.5. Under the most favorable conditions, the planet is never more than 30 degrees above the horizon, meaning that to look at Mercury you have to look through an awful lot of Earth's atmosphere at a point where seeing is guaranteed to be bad. So it's always hard to find Mercury in the sky. Even when you do get to see it telescopically, it is never larger than about 10 arcseconds in diameter. Scratch Mercury from our planetary observing list.

Next out from the Sun is another inferior planet, Venus. Venus is almost the same size as Earth, so it is much larger than Mercury. At its farthest point from Earth, called *superior conjunction* for inferior planets, Venus is about 10 arcseconds across, as large as Mercury is when it is closest to Earth. At its closest point, called *inferior conjunction*, Venus is almost a full arcminute in diameter. Venus also shows phases, like Mercury, but the mechanics of its orbit bring it farther from the Sun so it can rise to 47 degrees above the horizon. In addition, sunlight reflected from Venus' atmosphere causes it to outshine everything in the sky except the Sun and Moon. At between magnitudes −3.7 and −4.5, it is hard to miss this planet. Easy to find, the amateur must put Venus on his observing list.

Next we come to the Earth, a very difficult planet to view telescopically. Let's leave it off our list.

Outside the orbit of the Earth we come to the realm of the *superior planets* (Fig. 9-2). The first of these that we encounter is Mars. About half the size of Earth, Mars

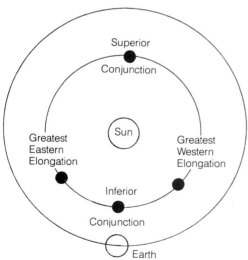

Fig. 9-1. The orbit of an inferior planet like Mercury or Venus. At inferior conjunction the planet will appear as a thin crescent. At superior conjunction it will appear as a full disk. At either elongation it will appear half-full.

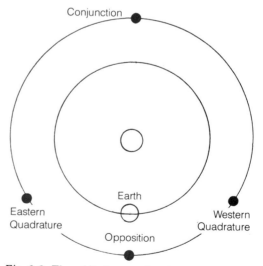

Fig. 9-2. The orbit of a superior planet such as Mars, Jupiter, or Saturn. At western quadrature the planet rises at midnight and sets at sunrise. At eastern quadrature the planet is on the meridian at sunset. At conjunction the planet is hidden behind the Sun.

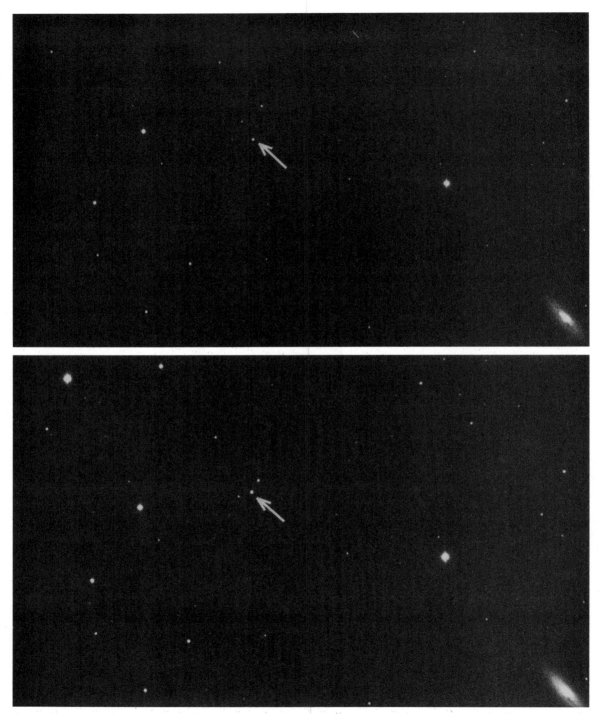

Fig. 9-3. The planet Pluto among the stars. (Courtesy Lick Observatory.)

moves around the Sun in a much more elliptical orbit than the Earth's. When it is opposite the Earth, at *conjunction*, Mars presents a very tiny disk that is less than 10 arcseconds across. When it is closest to us, in *opposition*, it can be as large as 25 arcseconds in diameter. Its size at opposition depends on where it is in its elliptical orbit so it can be as small as 14 arcseconds or as large as 25. Even though Mars can be quite small, it is still easy to find with the naked eye because its magnitude range is from about +2.0 to almost −3.0, and it has a reddish color. Mars is unique from all the planets in that it is the only planet that shows us a solid, undistorted surface. So again we have an easy-to-find planet with something to offer. On the list it goes.

Now we come to Jupiter, the largest planet in the Solar System. Easy to find, never below magnitude −2.0, and over 10 times the size of Earth, Jupiter always presents a grand telescopic sight with its cloud belts and satellites. This is a must-observe item for our list.

Finally, we come to the last of the planets known to science before Herschel added Uranus in 1781. Saturn is probably the most beautiful sight you will see in any telescope. Including its ring system, Saturn presents a target of between 30 and 40 arcseconds from tip to tip, the ball of the planet being slightly smaller. Why say anything more; Saturn is on our list, no questions asked!

Uranus is the first of the modern planets—those added after the invention of the telescope. Just at the naked-eye limit of magnitude 6, Uranus is a very large planet four times the size of the Earth. It is also a very distant planet and never appears bigger than 4 arcseconds across. Although it is relatively easy to find, it never shows anything more than a green, featureless disk, so Uranus is not for us.

Neptune is another of the distant children of the Sun. Although it is large, at four times the size of the Earth, Neptune is so far away that it only presents a blue-green disk less than 3 arcseconds across. Again, not for us.

Finally we come to Pluto, the farthest outpost of the Solar System. Pluto is so small and far away that it never gets brighter than magnitude 14 (Fig. 9-3). While you have at least a chance of glimpsing every other planet in the Solar System, Pluto is beyond the reach of almost all amateur instruments. Even if you do find it, it will show no disk. Definitely not for our observing list.

So that is the Solar System through the eyes of an amateur astronomer. It doesn't seem like much, just four planets: Venus, Mars, Jupiter, and Saturn. But believe me, you could spend a lifetime studying only one of them. Ah ha! I see doubt written across your face. Okay, let's see what an amateur can do with these planets.

CONSIDERATIONS FOR THE PLANETARY OBSERVER

A big factor in successfully observing the planets is your telescope. Each type of telescope has its own advantages and disadvantages concerning planetary observation. For years the measure of a good planetary instrument was its focal ratio. Longer focal ratio telescopes, f/12 up to f/15 and sometimes even longer, give the instrument a small, dark field of view that creates greater contrast between the planet under observation and the background sky. This allows for more detail to be seen on the planetary surface. Combined with a medium to large aperture and as unobstructed a light path as possible, long focal ratios and large apertures should allow for excellent detail resolution.

The best instrument for planetary observation has been the *refractor*. It nicely combines an unobstructed light path with long focal ratios. The problem with the venerable refractor is aperture. Most instruments over 4 inches in aperture are, to say the least, rather bulky. A 6-inch refractor (Fig. 9-4) with a focal ratio of f/15 is 90 inches long! It is rather difficult to transport, but not impossible. An 8-inch f/15, however, has a focal length of 120 inches. Try fitting that into your car! The mounting required for instruments this size border on the huge. For a stable image, the mount has to be steady, but try to make a mount smaller than Jack's beanstalk for a telescope with a tube almost 100 inches long.

There are ways around this problem of length. The simplest solution is to bend, or fold the light path by using high-quality optical "flats." Though some contrast is lost, a folded refractor with a 4- or 5-inch aperture and a focal ratio of f/15 or even f/20 becomes a manageable instrument on a mount that doesn't take an Olympic weightlifter to move.

On the flip side of the coin are the shorter focal length refractors. For years, a refractor shorter than f/15 was considered unsuited for planetary work. The chromatic aberration given off by a lens of that focal ratio guaranteed an image with pretty color rings around it. But technology does have its advantages. Computer-designed lens systems and special lens coatings allow for the construction of shorter, effective focal ratios.

The primary problem with refractors is their cost. A good 6-inch lens in a cell and tube but without a mount, can cost over $2,000. An 8- or 10-inch lens is out of sight for most amateurs.

Long focal lengths are also a problem with *reflectors*. The common newtonian design used today has a focal ratio of about f/5. According to many planetary specialists this is just short of the optimum focal ratio for good contrast in a planetary image. A newtonian that will be used for planetary work should have a focal ratio of f/6 or longer. But in astronomy and optics, just like everything else, there are exceptions. I have seen a 10-inch f/4 newtonian with absolutely superb optics give planetary images that rival a large refractor.

The perfect planetary reflector has a design used by Sir William Herschel, aptly called herschelian. In this design there are no obstructions in the light path because the observer looks directly over the edge of the tube into the primary mirror. Herschelians have huge focal ratios. The one in Fig. 9-5 is an f/27 with a 6-inch primary and a focal length of over 160 inches!

Again, the rule of thumb should be large aperture and long focal ratio. Once again the observer is faced with long, bulky tubes, and huge, weightlifter sized mounts. But reflectors are easier to adapt than refractors. Modern optical designs allow for the bending of the light path with special mirrors that eliminate obstructions and lengthen the focal ratio. The tri-schiefspiegler design (Fig. 9-6) enables a 6-inch primary with a focal ratio of f/20 to be 120 inches in focal length and to fit in a tube about 4 feet long.

There is a way to coax a longer focal length out of your newtonian. A 6-inch f/6 has a focal length of 36 inches and a large central obstruction, the secondary mirror, that reduces the image contrast. You can turn this reflecting telescope into a modified herschelian by adding an *aperture mask* (Fig. 9-7) to the system. An aperture mask is simply a piece of cardboard with a hole cut off-center. The piece of cardboard fits over the end of

Fig. 9-4. A 6-inch f/21 refractor with the objective constructed by its owner.

the telescope tube. The hole is placed so that it allows light to enter the instrument and bypass the secondary mirror and its supports. The hole is small, but the benefits are great.

For our 6-inch f/6 we will need a hole a bit larger than 2 inches in diameter. When this is placed in the system's light path, it changes the 6-inch f/6 to a 2-inch f/18 with a huge increase in the image contrast, which makes up for the loss in aperture.

Aperture masks are not limited to use on reflectors. Center the hole in the mask material and you create a mask for a refractor. You create the same effect by lengthening

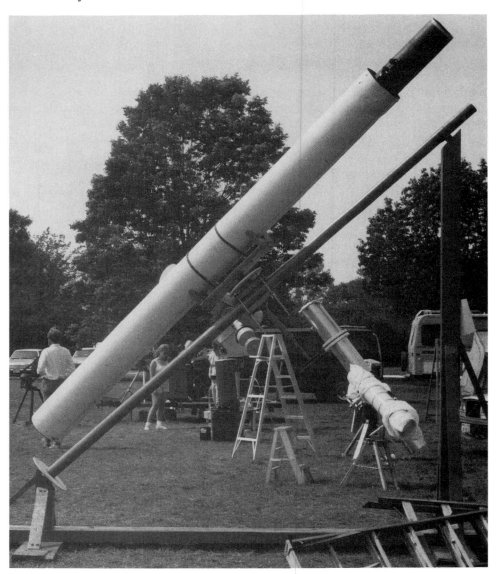

Fig. 9-5. A herschelian 6-inch reflector.

the focal ratio and making the aperture smaller. This can be particularly important with short focal ratio refractors because it helps to cut down on chromatic aberration.

An aperture mask also can help improve the overall performance of your telescope when you observe the planets. The light that reaches your eye has had a long, tough journey. Not only has it had to travel millions and maybe billions of miles through space, in the case of the planets, but it has also had to make its way through Earth's atmosphere. As the light moves through the atmosphere it runs into little pockets, or *cells*, of air, some warm and some cold. It is these cells that cause the image you finally see to shift and shimmer. The more of these cells through which the light that forms your image must pass, the more shimmer you will see, and the poorer the quality of the evening's overall seeing. If the light doesn't have to pass through too many of these cells, the image will be sharper and more stable.

A large aperture will gather light over a greater surface area, but the increased surface means you will look through more of these atmospheric cells. Using an aperture mask to reduce the light-gathering surface reduces the number of cells you are looking through and improves seeing.

But don't get me wrong! The aperture mask is to be used *only* during periods of poor seeing, not all the time. If you are observing and notice that the seeing clears up and the image sharpens for noticeable periods of time, take advantage of the full aperture

Fig. 9-6. A three-mirror 6-inch f/20 tri-schiefspiegler.

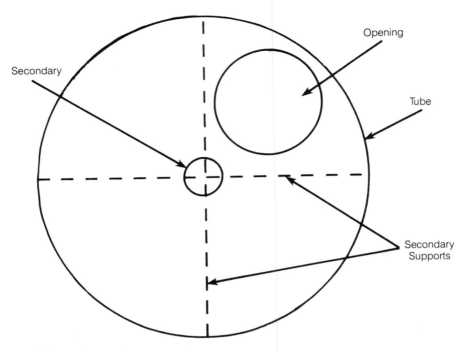

Fig. 9-7. The aperture mask.

and the light-gathering power and resolution of your primary. Use the aperture mask only as a secondary resort.

The final telescope design in popular use today is the schmidt-cassegrain (SCT). While many feel that these are the best telescopes from an overall standpoint, they do suffer problems when it comes to visual planetary observation. The primary problem is inherent in the design of the system. The mirror used in these instruments is usually a very fast f/2 in a very short tube assembly. This means that the light path is going to be reflected back to a secondary mirror a short distance away. For a 10-inch system this distance is about 2 feet. This requires the use of a large secondary mirror that blocks the light path. A large obstruction might not affect a stellar point source image, but it can only harm the definition of a bright, extended planetary image. The SCT optical system also causes a loss of contrast because the light enters the eyepiece assembly through a hole in the mirror. This problem can be avoided by using an effective system of baffles to reduce the extraneous light that enters the eyepiece.

All these problems might seem insurmountable, but they are not. A properly baffled SCT with a secondary obstruction of less than 20% will give good planetary images. Just make sure that your SCT meets these simple requirements.

But how big a telescope do you really need to observe the planets and do serious work? How big a telescope do you have? A 60mm refractor? An 80mm refractor? A 4.5-inch reflector? A 6-inch reflector? An 4- or 8-inch SCT? A 10-inch reflector? They are all fine and you can make serious planetary observations with any of them. You have to remember the limitations of your particular telescope and learn to observe within them.

The light grasp of a 10-inch reflector is fine until you turn it on the brilliant disk of Venus and find your best view of the planet comes when you use an aperture mask that reduces the aperture to a size of 3 inches. Certain projects require larger instruments, but for every one that does there are an equal number of projects that can be successfully undertaken with smaller-aperture instruments.

Okay, enough of this stuff. Let's go back to the planets!

OBSERVING VENUS

Venus is the easiest planet to find with the naked eye and the most difficult planet to study with the telescope. Sir John Herschel claimed this was because "The intense luster of its illuminated part dazzles the sight and exaggerates every imperfection of the telescope."

Point even a small instrument at Venus and you will see that Sir John was right! It almost hurts the eye to look at it. If you are using a refractor, the edges of the image will be ringed in yellow and violet colors caused by chromatic aberration. A reflector will show light scatter from every little pit in the mirror, and the vanes of the spider holding the secondary mirror will show sharp spikes of light. Venus is indeed a very difficult object to observe with the telescope!

Any telescope will show the phases of Venus, so following the planet throughout its *apparitions*, the period of time when it or any planet is visible, is an easy project (Fig. 9-8). As Venus moves in its orbit from behind and to the east of the Sun, it is in its *eastern*, or *evening*, *apparition* and will be visible in the western sky after sunset. As Venus moves higher into the western evening sky, it will eventually reach its highest point in the sky. There it is then said to be at its *greatest eastern elongation* at 47 degrees above the horizon. It will then appear to drop back toward the horizon until it disappears in the Sun's glare. As Venus moves between the Earth and the Sun, it moves into its *western*, or *morning*, *apparition* and will be visible in the eastern sky before sunrise. Again, the planet will appear to climb in the eastern sky until it reaches its highest point at 47 degrees above the horizon, when it will be at *greatest western elongation*. Once again, Venus will drop lower in the sky until it is lost in the Sun and the entire cycle begins again.

The best time to observe Venus is when the plant has a large angular diameter and a phase large enough to show a significant portion of the disk. This usually occurs midway between western elongation and superior conjunction, and again midway between superior conjunction and eastern elongation. During these periods Venus will appear between its half-full and full phases and have a diameter of about 16 arcseconds.

We never get to see the real surface of Venus because it is shrouded by a thick, cloudy atmosphere. What you see when you observe Venus is the top layer of the planet's atmosphere. It appears to be light yellow or yellowish-white in color to the observer and almost blindingly bright. There have been, however, tantalizing glimpses of something in the clouds over the years. In 1666 and 1667, the Italian observer Domenico Cassini saw a number of dark spots and one that was very bright.

Another Italian, Bianchini, reported spots on Venus many times during the 1726-1727 apparition. J. E. Bode, director of the Berlin Observatory, observed dark spots on the planet in 1788. In 1801 J. H. Schröter saw a dim, dusky streak. The greatest observer of the time, Sir William Herschel, was never able to make out anything more than some indistinct markings. In recent years, observations of dusky markings that look like spokes on a wheel have been seen in the clouds of Venus. Some correlation between these ghost markings and features on the true surface of the planet have been hinted at by photographs taken at the ultraviolet end of the spectrum.

If you want to accept the challenge and hunt for these elusive features, it would be to your advantage to invest in a set of filters to help cut the glare and at least partially pierce the clouds. Observing with filters allows you to *screen* the planet in a particular wavelength of light. Because of the increased contrast induced by use of the filters, features can become more apparent. Filters act to transmit certain wavelengths of light and turn away others (Fig. 9-9). A red filter will allow red light to pass through it while it turns away light at the blue end of the spectrum. A reddish-colored feature will appear bright

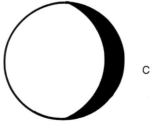

Fig. 9-8. Venus appears as a thin crescent of almost 1 arcminute in diameter at inferior conjunction. At elongation Venus will appear as half-full. Near superior conjunction it will appear as a gibbous phase almost full.

and stand out from the background, while blue features will blend into the disk and be difficult to distinguish.

A number of companies sell filter sets that screw onto the barrel of the eyepiece. These are useful, but every time you want to change filters you must remove the eyepiece, unscrew the filter, screw in a new filter, and then replace the eyepiece in the focuser. This can get to be a bit tedious. You can also buy filters by the sheet and hold them between the eyepiece and your eye. You don't need a large filter. Those best suited for planetary observing are the Kodak Wratten filters sold in sheets that are 3 inches square, and they are expensive. A number of small companies offer laminated cards that have up to eight filters mounted on them. You can take the filter sheets and mount individual colors in 2×2 mounts or even put a series of filter colors in a single mount. Table 9-1 gives you a breakdown of the filters in the Wratten line and their various transmission factors.

If you cannot find Wrattens there are other ways of obtaining filters. Theatrical supply companies often distribute sample packs of their gelatin lighting filters. Although these do not have the same transmission qualities of a Wratten and might include some strange color names like Tiger Red, they can be used. The important thing in using any filter is that you know its transmission range. These theatrical filters come with a breakdown of their wavelength transmissions so you can select the ones closest to the Wratten range.

Another filter that an observer will find useful is the variable density polarizing filter. It is made of two disks of polarizing material that are mounted together so that either filter element can be rotated independently. When the elements are rotated so that the axes are in phase, or parallel, light transmission is at its maximum. As you rotate either element, the resulting scene darkens until the elements' axes are at a 90-degree angle to each other and very little light is transmitted. A polarizing filter allows you to cut down on the glare from an object like Venus without affecting the white light transmissions of the image.

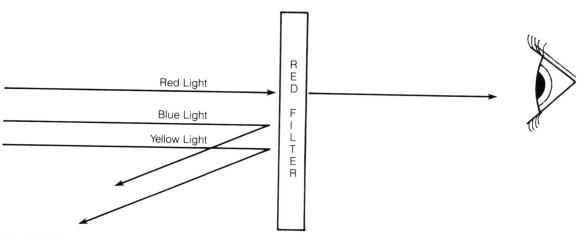

Fig. 9-9. Filters selectively admit certain wavelengths of light while reflecting others away.

Table 9-1. *Wratten Filters of Interest to the Amateur Astronomer*

Wratten Number	Filter Passes	Best Used To Observe
25	red	Moon, Mars, Venus
23A	red-orange	Mars
21	orange	Mars
15	deep yellow	Mars, Jupiter, Saturn
12	yellow	Mars, Jupiter, Saturn
8	light yellow	Mars, Jupiter, Saturn
57	yellow-green	Jupiter
58	green	Moon, Mars, Jupiter, Saturn
64	blue-green	Mars, Jupiter
80A	light blue	Mars, Jupiter, Saturn
82	light blue	Jupiter
82A	light blue	Jupiter, Saturn
38	light blue	Jupiter, Saturn
38A	blue	Mars, Jupiter, Saturn
47	deep blue-violet	Moon, Mars, Venus
30	light magenta	Mars

When hunting for the dusky features on the disk of Venus, you can use a violet (Wratten #47) or blue (Wratten #38A) filter to bring out these low-contrast features. Because the Wratten #47 filter is so dark, you should use it only if your instrument has sufficient light-gathering power to penetrate the filter. A 6-inch telescope or larger should be used in conjunction with this filter. If you have a telescope that is smaller than 6 inches in aperture, the Wratten #38A will give adequate penetration.

Another area of investigation for the amateur's telescope is a study of the terminator over the course of a number of apparitions. Because Venus exhibits phases like the Moon, it will show a terminator just like the Moon. On Venus this line separates the light from the dark portions of the planet (Fig. 9-10). The terminator, however, is not always a sharply defined line as it is on the Moon.

The terminator of Venus can exhibit a variety of conditions that you can study with a telescope. It can show a number of irregularities along all or part of its length. These might appear in white light and are best seen when using a polarizer. They also appear very well under red filtration and a Wratten #25 will bring them out. These irregularities might look like indentations that go from the dark, unlit side of the planet into the bright, sunlit portion. They also might appear as bulges that move from the bright to the dark hemispheres of the planet. Finally, they might appear as a wavy or serrated line alternating between the two hemispheres.

Related to terminator irregularities is a phenomenon known as the *schröter effect*. As the planet moves around the Sun it is expected to display certain phases at certain points along its orbit. When the planet is at or near either of its elongation points, it is expected to be at half phase, with the terminator theoretically being a straight line running from the north to the south pole.

The exact moment of the half phase is known as *dichotomy* and can be computed accurately to within seconds of when it is supposed to occur. The only problem is that

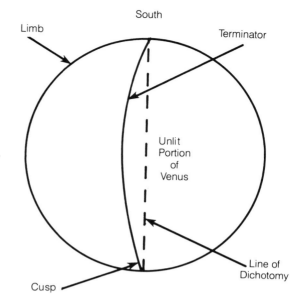

Fig. 9-10. Venus through a telescope.

the exact dichotomy is rarely, if ever, observed when it is supposed to take place. The difference between computed dichotomy and the observed half phase can vary by as much as a week. This is thought to be caused by refraction in the planet's atmosphere and the fact that while the limb of the planet appears sharp and distinct, the terminator shows up as a fuzzy line.

The tips of the terminator, called the *cusps* or horns, can also take on characteristics that can be observed with the telescope. Prominent bright areas or *cusp caps*, have been observed at or near the true cusps of the planet during its crescent phase. They might appear to take on irregular shapes and can be seen extending along the limb of the planet. The caps have also been observed to have a dark border around them, called *cusp bands*. The cusps of the planet have also been noted to be of unequal brilliance and size, with the south cusp being the more prominent of the two. It would appear that these cusp effects are similar to the polar caps on planets like Earth and Mars. In Venus' case, however, these are not permanent features. Data from spacecraft indicate that the planet does not have features analogous to polar caps.

A final feature of the cusps is their occasional extension so that they encompass the entire planet, giving it a *halo*. Caused by the refraction of sunlight through Venus' thick atmosphere, this halo effect is most often seen when Venus is near inferior conjunction and appears as a very thin crescent.

A final observable feature on Venus is its *ashen light*. First observed by the Italian astronomer Giovanni Riccioli in 1643, the source of the ashen light has continued to elude observers even after the arrival of space probes. The phenomenon appears as a faint glow on the dark side of the planet and seems similar to *earthshine*—light reflected from the Earth and seen on the dark side of the Moon. However, any similarities end there. Earthshine and the ashen light might look alike but they are caused by completely different mechanisms.

The best guess as to the cause of the ashen light is the suggestion that solar particles activate a mechanism in the atmosphere of Venus that creates a glow similar to Earth's aurora. It would seem to follow that during periods of great sunspot activity the ashen light should be visible. Although there is some evidence to show a connection between solar activity and the ashen light, there is also evidence that indicates the two are not connected. Observations by members of the British Astronomical Association (BAA) during the period of maximum solar activity in the late 1950s (1956–1958) included many sightings of the phenomenon. But observations during later periods of solar maximum do not show much in the way of observed ashen light. Also, data from spacecraft sent to Venus show no evidence for the light's existence. By the same token, however, they don't show any evidence to rule out its existence either.

The ashen light is a perfect example of negative data being important as positive data. If you observe Venus over the course of a number of apparitions, it is just as important to note when the ashen light was not seen as when it was. The only way to solve the mystery of the ashen light is to have as wide a database as possible from which to work. The only way to build such a base is to have as many observations, both negative and positive, as possible.

OBSERVING MARS

Outside of the Moon, the most fascinating object in the Solar System for the man on the street is Mars. Over the centuries a mystique has built up around the Red Planet to the point that when Orson Welles broadcast his adaptation of the *War of the Worlds* in 1938, well over 1 million people were willing to believe that we were being invaded by men from Mars!

A Tale of Two Orbits

Mars is the first superior planet we encounter. The two most important points in the superior planet's orbit with respect to Earth are *conjunction*, when the planet is opposite Earth and moving behind the Sun, and *opposition*, when the planet is closest to the Earth in its orbit. Although superior planets are farther from the Sun than Earth, they can still show phases. When a superior planet is near conjunction at the far end of its orbit, it will show its maximum phase, slightly past gibbous, and have the shape of a 10- or an 18-day-old Moon.

The orbital mechanics of a planet like Mars also dictate when we on Earth will get a good look at it. Mars, however, has a funny orbit. Although the Earth reaches opposition with Mars every two years, not every opposition is good for observing the planet because the orbit of Mars is not a perfect circle (Fig. 9-11). The average opposition, which takes place every 26 months, put the planets about 50 million miles apart and yields an apparent diameter for Mars of about 17 arcseconds. When Mars is closest to the Sun, at *perihelion*, the orbits of the two planets can be very close to each other, between 33 and 35 million miles. But because Earth moves faster in its orbit than the more distant Mars, it is not always there to greet the Red Planet. The most favorable views are afforded us when the planet's closest approach to Earth and the Sun coincide.

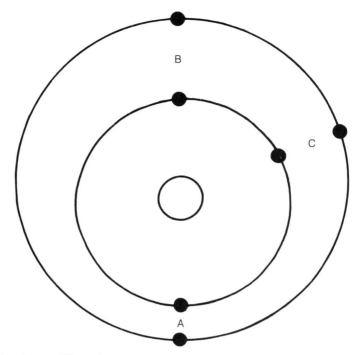

Fig. 9-11. Oppositions of Mars. At a perihelic opposition (a) Earth and Mars are close together. At an aphelic opposition (b) the two planets are much farther apart. The placement in (c) is when most oppositions occur.

These *perihelic* oppositions occur every 15 to 17 years and take place during the late summer. When these two orbital points coincide, the planet Mars can have a disk between 23 and 25 arcseconds in diameter. During perihelic oppositions, the southern hemisphere of the planet is always tilted toward Earth.

It is much more common for the Earth to catch up to Mars when the planet is at its farthest point from the Sun, or *aphelion*. This takes place every 9 to 12 years. During *aphelic* oppositions, the distance that separates the two planets is much greater, from 48 to 63 million miles. Because this distance is much greater than a perihelic opposition, the apparent diameter of the planet is much smaller. At oppositions like these, which occur during January and February, the size of the Martian disk is between 13 and 15 arcseconds in diameter. During aphelic oppositions, the northern hemisphere is always tilted toward Earth.

Telescopes and Mars

As with everything else in amateur astronomy, what you do is dependent on the equipment you use. Just because you have a small telescope doesn't mean you can't do meaningful work and make contributions to our ever-expanding body of knowledge about Mars. You just have to temper your enthusiasm with a bit of reality. To make the best use of your instrument, it is a good idea to get and use filters. Table 9-2 lists those recommended for use when observing Mars.

Wratten Number	Features Visible
47	Clouds high in atmosphere
38A/80A	Clouds at lower altitudes
	Limb hazes
	Polar hood
64	Low-lying fogs
58	Surface frosts
8/12/15	Dust storms
23A/25	Features on surface

Table 9-2. Reaching the Martian Surface with Filters

If you have a 60mm (2.4-inch) refractor you can't expect to see minute details on the Martian surface. During favorable perihelic oppositions, a 60mm can easily detect the polar cap and many of the dark surface features on the planet. Although you may not be able to pinpoint very small details, like individual craters on the Martian surface, you can easily see and record the shrinking or growth of the planet's polar caps as Mars approaches opposition. Using filters you can also follow the growth of clouds near the polar caps. This might not seem like much, but observations of the polar caps contribute a great deal to our understanding of Martian weather, something we are going to need a lot of information on before man sets foot on the planet.

Observing the planet Mars can be a challenge for the amateur. Because the usual apparent diameter of the planet is small, you will have to use your telescope near the limit of its useful magnification range in order to make the disk large enough to see detail. In doing so you will have to take particular care not to use a magnification that is so high it causes any surface features to blur and become fuzzy.

During the last favorable opposition in September 1988, I used my 80mm f/11.4 refractor to observe the planet. Using the guideline of 25D, or 25 times the diameter of the objective, that we discussed in chapter 4, I should have limited my maximum magnification to 75×, or 25 times 3. Images yielded by the 80mm and a 12mm orthoscopic eyepiece were sharp and detail on the planet's surface was quite visible. But rules were made to be broken. This applies especially when observing the planets. You can take advantage of particularly good seeing to raise the limit of your telescope to as much as 50 to 70 times the aperture. On a number of occasions during that apparition, I was able to use an 8mm RKE eyepiece with a 2× barlow lens to observe the planet. This yielded a magnification of 227× for my 80mm, or 75 times the aperture (75D). The exit pupil was small, but the results were satisfactory. I emphasize that when I tried this the seeing was excellent.

For most of the apparition I used an 18mm orthoscopic eyepiece with a 2× barlow, giving me 101×, or about 34D. Though still above the limit, remember that planets take magnification much better than other celestial targets. When you increase the magnification of an image, you are making it larger without increasing its overall brightness. This spreads the available light over a larger area and as a result the image becomes darker. Because the glare of a planetary image will wash out small details, the increased darkness of a magnified image can help bring out small surface details near the telescope's threshold of resolution.

The important thing to keep in mind is that you are trying to get the best resolution of detail possible from your telescope. There is nothing wrong with pushing your instrument to the limit when seeing permits. The key to that sentence is "when seeing permits." It is the stability of the atmosphere more than anything else that limits how far you can go. About the highest I would try to push the 80mm given excellent seeing is 227×.

A Martian Geography Lesson

Mars is the only planet whose surface features can be seen clearly from Earth. Most of the features on today's planetary charts (Fig 9-12) were mapped and named during the later 1800s by the Italian astronomer Giovanni Schiaparelli. Even though Mars is only half the size of Earth, these *albedo features*, the dark markings and desert areas, cover a surface area equal to the land surface of Earth because there are no bodies of water on Mars. The map of Mars is overlaid with a longitude and latitude grid to make it easy to identify the location of each albedo feature. You will notice that the longitude system for Mars begins at 0 degrees and wraps around the planet a full 360 degrees. This is to make it easier for us to find things on Mars when looking at the planet with a telescope.

Figure 9-13 shows the orientation of Mars when it is viewed through a telescope. South is at the top and north at the bottom, but notice the names of the east and west limbs of the planet have been changed. East is now the *preceding limb* (p) and west is the *following limb* (f). The preceding limb is the side of the planet moving into evening; the following limb is the portion of the planet under the morning Sun. Features on the planet will move west to east, from preceding to following limb, through the course of a Martian day called a *sol*.

The terminator, also called the *phase defect*, separates the illuminated and unilluminated portions of the planet's disk. The portion of Mars illuminated by the Sun is never less than 85%. When the planet is seen before opposition, the terminator denotes the line of sunset on the planet and appears on the evening (p) limb. After opposition it marks the line of sunrise and appears on the morning (f) limb.

There is another line drawn on the disk of Mars labeled the *central meridian* (CM) of the planet. It is an imaginary line that passes through both poles at a right angle to the equator and is equivalent to the prime meridian on Earth. The central meridian is used to tell us the longitude of the feature passing across the center of the planet's disk. If the planet is past opposition and a phase defect is present, the CM will appear off center on the disk.

Because a day on Mars is longer than on Earth at 24 hours 37 minutes, the planet will rotate through 350 degrees of longitude for each 24-hour Earth day. The difference of 37 minutes causes markings on the planet to appear to move backwards by 10 degrees longitude from night to night. So if you observe a feature on the following limb of the planet one night, the following night it will be around the other side of the planet. It will be five or six weeks before you can see that feature again.

Information on what longitude is on the CM for any given date is found in such references as the RCAS's *Observer's Handbook*, the *Astronomical Ephemeris*, and the *ALPO Solar System Ephemeris*, which lists information on all the objects visible in the Solar System. So if the ephemeris you are using gives a CM longitude of 91 degrees for a given

Fig. 9-12. Mars map. (Courtesy ALPO.)

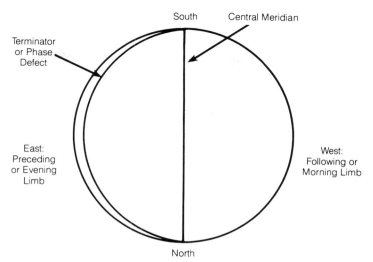

Fig. 9-13. *Mars through a telescope.*

date, you will know that the albedo feature Solis Lacus is on or near the center of the disk.

Another piece of information it is helpful for an observer of Mars to know is the time of year on the planet. This is given in an ephemeris as the longitude of the Sun (Ls) as seen from Mars. These longitudes denote the season on the planet and may seem tough to understand at first. The Ls for a particular date is figured for the northern hemisphere of Mars. Ls = 0 degrees indicates the point for the beginning of Martian spring. Ls = 90 degrees is the beginning of summer. Ls = 180 degrees is the beginning of autumn, and Ls = 270 degrees is the beginning of the Martian winter. Perihelion always comes at Ls = 250 degrees and aphelion at Ls = 70 degrees. Remember, these apply to the northern hemisphere of the planet, so the seasons in the southern hemisphere are exactly opposite, just like on Earth. If your ephemeris tells you that the longitude of the Sun (Ls) for the date of your observation is 200 degrees and it is an aphelic opposition, you know that it is early autumn. For the same Ls during a perihelic opposition, it is early spring, in the Martian Southern hemisphere.

Martian Polar Caps

The first thing you see when you look at Mars with a telescope is the polar cap (Fig. 9-14). During aphelic oppositions the north polar cap (NPC) is prominent. During perihelic oppositions the south polar cap (SPC) is visible. Made up of frozen carbon dioxide and water ice, the polar caps are the prime movers of Martian weather. Clouds that form over the surface of the planet are fed by the evaporation or *sublimation* of the caps, particularly the larger SPC.

Because the polar caps are so prominent and their growth and retreat are so important to science, monitoring them on a nightly basis is a perfect project for the amateur with a small telescope. A feature to look for when Mars presents its NPC is the *polar hood*. Although the SPC can also have a polar hood, the one that forms over the NPC is

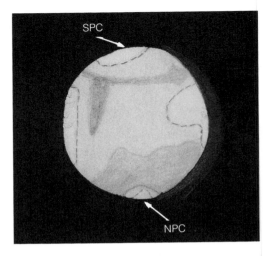

Fig. 9-14. Mars and its polar caps. (Courtesy Dan Troiani.)

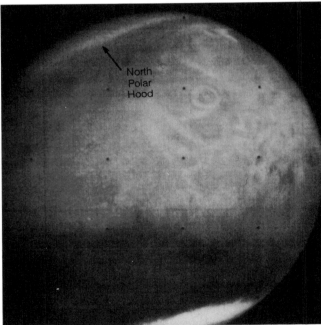

Fig. 9-15. Martian polar hood as seen from the Mariner 6 *spacecraft. (Courtesy NASA.)*

more prominent (Fig. 9-15). The north polar hood (NPH) forms during late spring on Mars (Ls = 70 – 80 degrees) and is associated with what scientists refer to as an *aphelic chill*, during which time the retreat of the NPC suddenly stops. Polar hoods can appear bright (Fig. 9-16), so bright that you could confuse the hood for the polar cap itself. Now is when filters come in handy. To determine if you are seeing a polar hood or the polar cap itself, look at the feature first with a blue (Wratten #38A) filter. If the feature is a cloud or the hood, it will appear bright in the blue light transmitted by the filter. Quickly switching to a red (Wratten #25) will cause the feature, it if is the hood, to disappear.

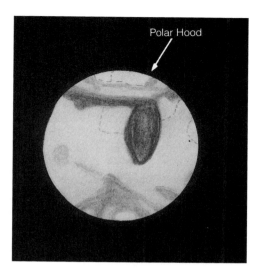

Fig. 9-16. Martian polar hood as seen through a 10-inch telescope. (Courtesy Dan Troiani.)

With larger instruments over 8 inches in aperture, you may be able to follow the shrinking of the polar cap and watch for the appearance of features that the retreating polar cap uncovers. Features to watch for in the NPC include Rima Tenuis and small projections of ice that ring the NPC during spring and summer. These are known as *Ierne* at longitude 150 degrees, *Lemuria* at longitude 60 degrees, and *Cecropia* at longitude 270 degrees. These features are best seen using red and green (Wratten #58) filters. Features in the SPC to watch for include the ice projections called Novus Mons at longitude 310 to 330 degrees, Thyles Collis at longitude 230 degrees, and Thyles Mons at longitude 180 degrees. These features are best observed during mid-spring (Ls = 230 – 240 degrees) and disappear with the full arrival of summer in the southern hemisphere. You also may notice that during summer the SPC is divided into two sections separated by the Mountains of Mitchell, which are not mountains but another ice remnant of the fully developed polar cap.

Martian Weather

For a planet with so little available liquids to condense and form clouds, Mars is very active meteorologically. During favorable oppositions even the owners of small telescopes can follow the development of clouds and dust storms across the surface of the planet. Filters are very important for the observer wishing to study the Martian clouds because they can tell the observer just how high a cloud is in the planet's atmosphere. Violet (Wratten #47) filters penetrate the Martian atmosphere the least. This filter is useful in detecting high-altitude cloud features. Blue filters (Wratten #80A, 38, and 38A) penetrate farther into the atmosphere and enhance clouds at lower levels than the violet filter. Blue-green filters (Wratten #64) emphasize features low to the ground but still elevated. A green (Wratten #58) filter will brighten features like frosts that are on the surface of the planet.

The easiest clouds to detect on Mars are those that form along the slopes of highly elevated areas like the huge volcano Olympus Mons. These are called *orographic clouds*

(Fig. 9-17), and they appear as single, isolated bright spots in the atmosphere when Mars is viewed without a filter. They are best seen in the spring and summer when the polar cap is shrinking. Their appearance through blue and violet filters indicate that they form and move through the upper levels of the planet's atmosphere. The wispy clouds that form near the peaks of mountains and the tops of tall buildings like the Sears Tower are Earth's counterparts to orographic clouds on Mars.

Another target for the small telescope is the detection of *limb hazes* along the terminator (Fig. 9-18). These atmospheric features do not rotate with the planet and might be localized cloud formations caused by rapid changes in temperature that occur at Martian nightfall or before sunrise. They do extend high into the atmosphere, as is indicated by their brightness when seen through blue and violet filters, and they can be harbingers of upcoming weather on the planet. The color of these limb hazes can also indicate if a dust storm has begun on the side of the planet facing away from Earth.

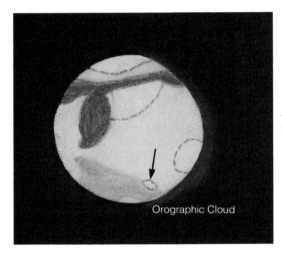

Orographic Cloud

Fig. 9-17. Orographic clouds using 10-inch telescope and a blue (Wratten #38A) filter. (Courtesy Dan Troiani.)

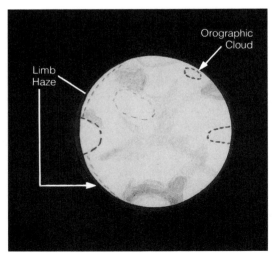

Orographic Cloud

Limb Haze

Fig. 9-18. Limb haze and an orographic cloud. (Courtesy Dan Troiani.)

Because Mars is subject to rapid and drastic temperature variations, *frosts* and *fogs* appear in low-lying surface areas. Visible as bright spots or patches, usually in the planet's desert regions, they are deposits of frozen carbon dioxide and water-ice laid down during the cold Martian night. As these frosts are heated by the Sun they create fogs. By local noon they have usually disappeared. Because the frost is deposited directly on the surface, it will appear bright when viewed through a yellow or green filter. The fogs hang above the surface and are best seen with a blue-green filter.

The bane of Mars observers has always been dust storms. Growing swiftly from the desert basins, these storms can quickly blanket large areas, or even the entire planet, in a matter of days. Since 1877, 40 dust storms of varying intensity have interrupted visual observations from Earth. Six of these, four since 1970, have been so severe that all sight of the planet's surface was lost. Dust storms are caused by winds that would make the most severe earthly tornado seem like a spring zephyr. Racing across the Martian landscape at nearly 300 miles an hour, the dust storm throws sand and surface material miles into the atmosphere. While they arise suddenly and can stop abruptly, the materials they have tossed into the atmosphere may take weeks or months to filter back down to the surface.

Yellow clouds, usually seen in the area of Solis Lacus, Chryse, and other basin areas, are often precursors of these dust storms. They show up best with yellow filters. Orange (Wratten #21) or red filters will show the observer the cloud's boundaries. Although a dust storm can start at any time during the Martian year, the planet-wide storms observed since 1970 have all begun when the longitude of the Sun was between 200 and 360 degrees—early to late summer for the southern apparitions, and early to late winter for the north.

Martian Surface Features

After a first-time observer has seen the polar caps, the next things to catch the eye are patches of green that appear across the disk of Mars. These are the raised areas of the planet. They appear quite distinct from the depressed, orange areas that define the planet's deserts and basins. These albedo features seem to darken with the retreat of the polar cap. The changes take place, not because of the growth of vegetation spurred by the water from the poles as was once thought, but because the areas surrounding the dark features become lighter as dust and particles suspended in the atmosphere during the winter filter down to the surface. The darkening of the albedo features is nothing more than an effect of contrast.

The albedo features can undergo other changes throughout the course of a Martian year. The region of Hellas, a huge impact basin, undergoes weather-related changes during the southern hemisphere's late autumn and early winter and should be watched closely. A gray hood can be seen to cover the area until the region moves into spring. As the planet moves into southern spring and the overall weather activity of the southern hemisphere increases with the retreat of the SPC, Hellas will undergo changes such as the growth of frost patches in the southern area of the basin, and cloud activity moving up from the SPC. Later in spring, the region should clear of ice and frost and its floor will appear to darken in color. As summer gets closer, the basin shows signs of dust storms, which should be closely monitored by the observer. The last three dust storms, two during the 1986 opposition and one during the 1988 opposition, began in the Hellas basin.

Changes like those in Hellas are seasonal, last only a short time, and can be predicted with some accuracy. Areas of the planet also can undergo longer-lasting *secular* changes that can occur abruptly and alter the appearance of an albedo feature for a long time.

Solis Lacus and Syrtis Major are particularly prominent features on the planet that can be easily monitored. When viewed in a telescope, Solis Lacus appears as a dark feature surrounded by a lighter desert area. It looks like a giant eye staring out from the surface of Mars, which is how it got its common name, the Eye of Mars. Observers of the area report that the Eye of Mars can undergo dramatic changes in its shape, size, and brightness. During recent oppositions it has been seen to grow larger and darker while very delicate, dark lines have appeared in the surrounding areas.

Syrtis Major (Fig. 9-19) is probably the most easily recognized feature on the planet. First seen by Christian Huygens in 1659, observations of the feature made during the next few years allowed early astronomers to determine that the planet rotated on its axis in 24 hours 40 minutes, a difference of 3 minutes from today's rate as determined by the *Viking* orbiters. Syrtis appears as a dark, triangular feature surrounded by lighter desert areas. Seasonal changes are predicted by the ALPO Mars Section for the peninsula-like feature each time the planet passes perihelion, when it seems to narrow, and after aphelion, when a general expansion of Syrtis takes place. The area has also been subject to a number of secular changes over past apparitions. During the past 15 oppositions, Syrtis Major has appeared to become smaller in size, losing the sharp "parrot-beak" at its northern tip, and growing more narrow across its center. This could indicate the expansion of the desert areas, which would ultimately change the entire face of Mars.

Occasionally visible in larger telescopes are the spidery, almost gossamer-like markings of the famous canals of Mars. First observed by Schiaparelli in 1877, these features have stirred up more controversy than any other feature on Mars. On the maps Schiaparelli crafted during the 1877 opposition, he drew a number of dark, almost straight lines that he labeled *canali*. To Schiaparelli, who wrote in Italian, this was the

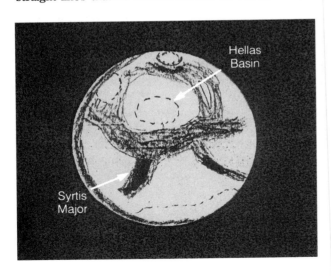

Fig. 9-19. The Hellas basin and Syrtis Major. (Courtesy Lee Keith.)

word for *channel* and was meant to indicate a long, almost straight feature that had nothing to do with intelligent beings. The problem began when other astronomers who did not speak Italian wrongly translated canali to canal. When the public heard about this they went nuts! There was somebody on Mars digging canals and they had to be intelligent! The debate raged for 17 years.

Then an East Coast amateur named Percival Lowell entered the fray. In 1894 Lowell erected an observatory in Flagstaff, Arizona, and obtained a 24-inch refractor from the firm of Alvan Clark & Sons, said to be one of their finest optical efforts. From this observatory, which still bears his name, Lowell set out to prove that Schiaparelli's canali were really the work of a race of master builders. Lowell observed over 400 canals crisscrossing the planet in a complex network. His efforts to popularize his theories met with some success. He was even able to convince some prominent astronomers that he was right about the canals.

The arrival of spacecraft on Mars settled the argument that raged through astronomical and popular literature for almost 100 years. The *Mariners* found no canals and the *Viking* landers were not met by members of Lowell's race of master builders. But, strangely, the spacecraft did confirm that Lowell had seen something. Some of his canals appeared to be chains of dark-floored craters, and some followed close to the alignment of dark-surface patches on the planet.

The discovery of Valley Marineris (Fig. 9-20) silenced many for a time. The "Valley of the Mariners" was discovered by the *Mariner* missions to Mars in the late 1960s. The valley is immense. It stretches over 3,000 miles and is 150 miles wide and 4 miles deep. It is a huge canyon that crosses one-quarter of the planet's surface. It is often referred to as the Grand Canyon of Mars. Theories abound as to exactly how the valley was formed. Some scientists say it was cut by furious water action at some time in the planet's distant past. Others insist that it is a huge fault line in the planet's surface that collapsed after a major volcanic eruption.

If you take your telescope out during the next favorable opposition and seek out the feature Tithonius Lacus (Fig. 9-21), you might notice a network of bright patches of fog and mist in the area. That is fog and mist gathered in the Valley of the Mariners and is easily visible from Earth, seeing permitting. The valley is evidence that of the two, Schiaparelli and Lowell, Schiaparelli was closer to the truth. The valley is a canali in the true sense of the word: a channel.

But what did Lowell see? Did he waste his fortune, time, and reputation pursuing an optical illusion? In a sense he did because there are a number of optical effects that can cause you to see canals. The eye can blend separate details into a single continuous image that will appear to be a straight line. The action is very similar to the way letters are created by a dot-matrix printer. From a distance each letter appears to be a solid, continuous impression on the paper, but up close it is easy to see the individual dots that combine to create the letter. This blending of discrete images is seen by numerous observers.

The prominent observer E. E. Barnard, possessor of the best eyes in astronomy at the time, was able to discern several craters that looked like dark patches on the Martian surface during the opposition of 1892. He used the 36-inch refractor at Lick Observatory. Are the craters that dot the surface of the planet the reality behind Lowell's canals? Go out and have a look at Mars. Then make up your mind.

Fig. 9-20. Valley Marineris as seen from 225,000 miles above the planet by Viking I. *(Courtesy NASA.)*

If you approach the observation of Mars thinking only of seeing the legendary "canals," you are going to be disappointed. You cannot expect to see this effect every time you turn your telescope on Mars. It takes extremely stable seeing and patience to discern the subtle markings. At an observing session during the recent excellent 1988 opposition, a number of observers commented that they had seen the "canals" using a 10-inch, f/20 tri-schiefspiegler and a 10-inch, f/9 newtonian. These observers, including ALPO Assistant Mars Recorder Dan Troiani, were blessed with spectacular seeing conditions (Fig. 9-22). The air was saturated with moisture which placed some estimates of the seeing conditions at 10, the highest level.

Drawing Mars

One of the real joys of amateur astronomy is sharing the results of your observations with your family, friends, and other observers. Your observing logs, journals, diaries, or

Fig. 9-21. Tithonius Lacus, a portion of the valley Marineris, as seen from the Mariner 9 *orbiter from 1,225 miles above the planet. (Courtesy NASA.)*

Canal Effect

Fig. 9-22. The subtle canal effect takes excellent seeing to detect. Some of the effect is caused by frosts and fogs lying in the valley Marineris. (Courtesy Dan Troiani.)

whatever you want to call them are one way to do this. In the case of an object like Mars, however, they require you to reduce visual impressions to words and that is difficult to do. There is often so much detail visible on the planet that words cannot convey their beauty and power. As someone once said, "A picture is worth a thousand words." That person must have been looking at Mars!

Making astronomical drawings is not as difficult as it may seem. When drawing a planet like Mars, you are not trying to rival Van Gogh. Very few people have that talent. What you are doing is making a permanent record of your visual impressions of the planet's surface over a short period of time. "You are not making a work of art," says Dan Troiani, "you're making a work of science." In chapter 8, we learned a simple way to make accurate renditions of lunar features using outline charts. When you draw the surface of Mars you are dealing with an ever-changing subject, so you won't have the luxury of using a static start provided by an outline chart.

The first thing to do is draw a blank circle. It can be drawn on a page in your journal, a 4 × 6 card, or a sheet of paper as long as it meets one requirement: it must be the size required by the organization to whom you will eventually submit the drawing. Most groups require you to use a blank disk between 40 and 50 millimeters across. ALPO, the logical target for any drawing made by a planetary observer, can supply you with forms for recording images on Mars and the other planets. In the case of Mars, ALPO uses a blank disk 42mm in diameter (Fig. 9-23).

Using a compass or a template of the proper size draw a few disks on file cards for use at the telescope. You will also need a few pencils, sharpened and of varying hardness, a timepiece that keeps accurate time, a red-shielded light, and something to work on such as a simple clipboard. Before you head outside, take down your ephemeris and check the night's phase defect (terminator). Take a couple of minutes to draw it in on a few of the blanks. If the planet shows a phase defect and you don't record it on the forms, they will be worthless because every detail you record will be in the wrong place. Now go out to the telescope!

Begin by getting comfortable at the eyepiece. If you are using a refractor, pull up a chair and sit down if you can. If you are using a reflector, get as comfortable as possible at the eyepiece. One advantage of long focal-length newtonians is that you will probably have to use a ladder to reach the eyepiece, and the ladder will provide you with a convenient ledge on which to rest your supplies. Select an eyepiece that allows you to see as much detail as possible without fuzzing out the planet. Take your time and give the image

ALPO MARS OBSERVATION - VISUAL - PHOTOMETRIC - PHOTOGRAPHIC - NO. _____
****

Fold on Line

DATE (by UT)_____.   OBSERVING NOTES (below and over)

L$_s$_____°   D$_e$_____°.

TIME (UT)_____—_____ CM_____°_____°

TELESCOPE_____.

MAGNIFICATION_____X,_____X.

FILTERS_____.

SEEING (0-10 scale)_____TRANS. (0-6^m)_____

OBSERVER_____

ADDRESS_____

_____

OBSERVING STATION_____

_____

Fig. 9-23. ALPO Mars observation form.

a chance to settle down. Take a few minutes to reacquaint yourself with the visible features. Try various filters to see what effect they have on seeing. Watch the focus of the image. Changes in seeing may require you to refocus the image from time to time.

Begin your actual drawing session by noting on your form the date and time that you start. Because Mars rotates at about the same rate as Earth you will have plenty of time to get all the details on your sketch. Estimate the overall seeing on a scale of 0 to 10, and the transparency on a scale of 0 to 6. Draw in the polar caps, south at the top and north at the bottom, depending on whether one or both are visible (Fig. 9-24). Go over the telescopic image and study it in white light (with no filter) and then with a yellow or red filter to identify surface features.

Start at the center of the disk and draw an outline of the major features (Fig. 9-25). These steps should take about 15 minutes to complete. Now grab your filters and start to add details to those outline forms. Darken the familiar features and indicate surface details and atmospheric features with dotted outlines for bright features like frost, and dashed outlines for clouds and hazes, as in Fig. 9-26.

The entire process of drawing the Martian disk should take from 40 to 50 minutes. You can't take any longer because the details have to be recorded before they rotate around the limb of the planet. It's not a bad idea to get the detail on the preceding limb down first because that will be the first area to rotate out of view. When you have completed the drawing, give the planet a final once over and note the time of completion on your form.

Fig. 9-24. Drawing Mars I.   Fig. 9-25. Drawing Mars II.   Fig. 9-26. Drawing Mars III.

Back inside take a few minutes and add to your observing form information on the telescope and magnifications you used during the drawing session; the filters you used—if they are not Wrattens try to include their transmission factors—and the longitude of the CM at the beginning and end of the session as obtained from an ephemeris. While you have the ephemeris open, scribble down the Ls for the date of the drawings. If you are going to submit this to ALPO, transfer the information from your journal or record card to their report form.

OBSERVING JUPITER

If you liked the show that Mars puts on with changing weather patterns and a wide variety of surface detail, you are going to love Jupiter! You can contribute a lot of important work to the study of Mars with a small telescope, but Jupiter can be a veritable treasure house for the amateur.

The biggest advantage that Jupiter gives the amateur with even a 60mm telescope is its size. At its smallest, Jupiter is 30 arcseconds in diameter, larger than Mars at its best. At its largest apparent diameter, 50 arcseconds, it is absolutely spectacular. Its magnitude range, between -1.3 and -2.7 with an average at opposition of -2.3, makes it easy to find with the naked eye. Unlike the orbit of Mars, with its great eccentricity, the nearly circular orbit of Jupiter makes it easily visible for 10 months our of every 13. Like Mars, Jupiter will show a very slight phase defect when it is near conjunction. For all intents and purposes, Jupiter was tailor-made for the amateur astronomer.

Unlike Mars, Jupiter does not have a visible, solid surface. A great ball of gas, the surface we see is only the top of Jupiter's atmosphere. Constantly in motion, the Jovian atmosphere rotates at two different speeds. The equatorial zone and its accompanying belts, System I, rotates in 9 hours 50 minutes, while the rest of the planet, System II, takes 5 minutes longer to circle the planet. The features that appear across this rapidly moving stage are the tops of clouds and storm systems driven by winds moving at greater than hurricane force.

Even a 60mm refractor can show an amazing amount of detail on Jupiter (Fig. 9-27). At $100\times$ you will easily see its squashed appearance due to its rapid rotation and the two, or more, major belts of the planet. The Great Red Spot (GRS) is also visible. If it is in a period of dormancy, you might be able to detect the great cavity in which it resides, the Great Red Spot Hollow (GRSH).

Details on the planet are not the only attractions. With its retinue of four bright satellites, the Jovian system offers the user of a small telescope a variety of satellite phenomena to observe. As they move in their orbits around Jupiter, the satellites cross the face of the planet (transit), move behind the planet (occultation), and move through the shadow of the planet (eclipse). In addition, the shadows of the satellites can be seen transitting the planet. This is a very busy system to observe.

For the observer with a larger telescope, there is all this plus the fine details to be seen in the belts of Jupiter. Watching and recording the movements of tiny light and dark spots and splotches across the surface of this giant planet can give you a lifetime of pleasure and the opportunity for scientific contribution.

Transits of the Central Meridian

Geography on Jupiter is defined by the systems of light areas, or *zones*, and dark areas, or *belts*, that circle the planet (Fig. 9-28). The closest thing to a permanent feature on the planet is the GRS, which has been around for over 300 years, and three white oval spots visible in the south temperate belt (STB) since 1939.

This lack of permanent features and the planet's rapid rotation make it difficult to determine the location of any features that appear on Jupiter. As with Mars, the primary resource for locating, and relocating, a feature on Jupiter is the longitude of the feature. How can you determine the longitude of something that is moving against a background that has nothing with which to relate its position? This is not the impossible task that it would seem. All that is needed is a telescope, from 60mm on up, a pencil, a notebook, an accurate watch, a calculator, and a copy of one of this year's ephemerises. The reduction of your results are so simple that the calculator is optional!

The basic work has already been done for you by the wonderful people who publish your ephemeris. All you have to do is note the time the feature passes the planet's CM. The CM of Jupiter is the same as it was on Mars, an imaginary line running from pole to pole at a right angle to the equator. On Jupiter, determining the CM is a bit easier than for Mars because Jupiter is not tilted toward or away from Earth by more than a few degrees. You can eyeball the CM or purchase a *micrometer reticle eyepiece* (MRE).

An MRE has two or more finely scribed lines at right angles to each other etched on a glass plate that is placed at the focal plane of the eyepiece. To use it, simply position one of the lines parallel to Jupiter's equator, which is roughly in the middle of the equatorial belt (EB). The other line defines the planet's CM. When the feature you are observing crosses that line, either determined by sight or the reticle, it is said to transit the CM. Because Jupiter rotates so rapidly, it is best to mark the time of this transit at three points, then average the three. The first timing is made when the leading, or preceding (P) edge of the feature crosses the CM. The second is made when the approximate cen-

Fig. 9-27. *A page from the observing log of Barbara Lux of McKeesport, Pennsylvania, showing Jupiter. Jupiter can yield a tremendous amount of detail even in a small telescope. (Courtesy Barbara Lux.)*

ter (C) crosses, and the final timing is made when the following (F) edge reaches the CM.

An ephemeris like the *Astronomical Almanac* lists the line of longitude that is on the CM of Jupiter at 0 hours UT for each day of the year. If you have not made your timings in UT, convert them by using the table in chapter 6. Multiply the number of hours since 0 hour UT by 36.58 for features in System I or 36.26 for those in System II. Then find the number of minutes that enter into the timing from Table 9-3 and add it all together. The result will give you an accurate determination of the feature's longitude to within 1 degree.

For example, let's say you see a dark spot in the EB on August 1, 1988, and time its three points of transit as follows:

Preceding edge (P) . . . . 5:27 UT
Center (C) . . . . 5:30 UT
Following edge (F) . . . . 5:34 UT
Average for transit . . . . 5:30 UT

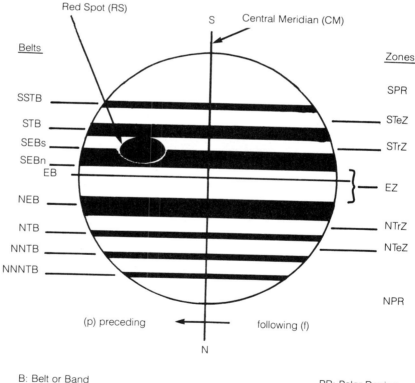

Fig. 9-28. *Jupiter's belts and zones.*

Table 9-3. Adjustments for Jupiter CM Transits

Minutes	System I	System II
1	0.6	0.6
2	1.2	1.2
3	1.8	1.8
4	2.4	2.4
5	3.0	3.0
6	3.7	3.6
7	4.3	4.2
8	4.9	4.8
9	5.5	5.4
10	6.1	6.0
60 (1 hour)	36.58	36.26

Source: Philip Budine, ALPO Jupiter Recorder.

From the ephemeris you determine that the longitude on the CM for 0 hours UT on that date is 75.3 degrees. Multiplying 5 by 36.58 gives you 182.9. Table 9-3 tells you that the correction for 10 minutes in System I is 6.1 degrees, so multiply this by 3 (3 × 10 = 30 minutes) to obtain 18.3. Add the two results to the listing from the ephemeris and you have a longitude of 75.3 + 182.9 + 18.3 = 276.5 degrees.

If you cannot time the transit of the complete feature because it has passed the CM by the time you see it, make sure to note the section of the feature from which the timing is made (Fig. 9-29).

In addition to taking actual timings of the transit of a Jovian feature, you have to be able to describe it. Features on Jupiter are either bright and white (W) or dark (D). They can be further described by using the classifications devised by Phillip Budine, former ALPO Jupiter Recorder and present Assistant Jupiter Recorder, in Table 9-4. These include six dark features (Fig. 9-30) and six bright features (Fig. 9-31).

Jovian Features

The oldest recorded feature on Jupiter is the GRS (Fig. 9-32). First observed by the English scientist Robert Hooke in 1664, it is a huge storm spinning through Jupiter's southern hemisphere. Like a hurricane in the southern hemisphere of Earth, the GRS spins counterclockwise and takes six days to rotate around its center. The GRS is lodged in a huge bay, the GRSH, that extends over the southern equatorial belt-south (SEBs) and spills into the southern tropical zone (STrZ). Winds in the belts and zones of Jupiter travel in opposite directions, adding rotation and extending the life of a feature trapped between them like the GRS.

The color of the GRS can vary from pale yellow-gray to deep red. The intensity of the colors can vary greatly from one apparition to the next. During the 1960s and early 1970s, the GRS was a prominent red-orange color, but it began to fade in 1976. Observers in the 1980s have noticed an increase in the prominence of the GRS, suggesting that the deep-red color, which has been its trademark, may be on the verge of returning. If the GRS is not visible, an observer may be able to detect its bay, the GRSH. The hollow is a much paler feature. If the observer looks carefully, the outline of the GRS can be made out inside the hollow.

TRANSIT OBSERVATIONS

PLANET_____   OBSERVER_____
YEAR_____   TELESCOPE_____
                                   LOCATION_____

No.	Date (UT)	U.T.	Object	Location	CM 1	CM 2	Remarks

ASSOCIATION OF LUNAR AND PLANETARY OBSERVERS

Fig. 9-29. ALPO form for recording transits of Jupiter's central meridian.

Table 9-4. Jovian Description Classifications

Dark (D) Features

Projections: An area extending from the edge of a belt.
Rod: An elongation running parallel to the equator.
Festoon: A filament crossing a zone.
Loop festoon: A curved festoon extending from a projection.
Column: A vertical form penetrating from a belt to a zone.
Disturbance: Any large, sharply defined area.

Bright or White (W) Features

Oval: A bright, well-defined round area.
Nodule: A feature similar to but smaller than an oval.
Notch: A semi-circular indentation on the edge of a belt.
Bay: A feature similar to but larger than a notch.
Rift: A long, thin streak along the interior of a belt.
Streak: An elongated white spot.

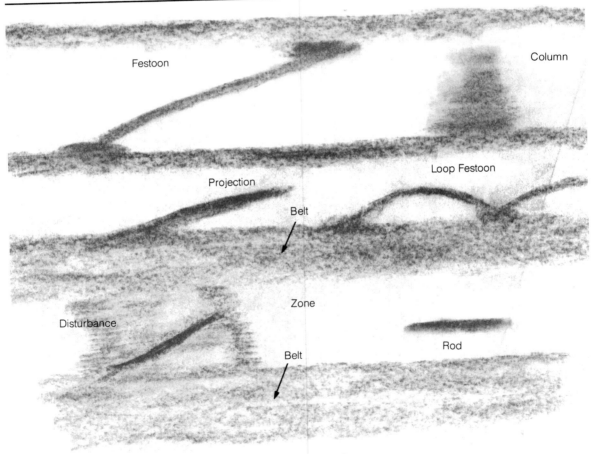

Fig. 9-30. Jovian dark features.

Fig. 9-31. Jovian light features.

Fig. 9-32. Jupiter's Great Red Spot as seen with an
8-inch telescope. (Courtesy Daniel Joyce.)

White Ovals

Fig. 9-33. The white ovals on Jupiter. (Courtesy Dan Troiani.)

Because of the size and prominence of the GRS, a study of it and its movement across the planet is ideal for the amateur with a small telescope. Transit timings taken at all three points—preceding, center, and following—can provide valuable information for determining how the storm has lasted over 300 years.

Other features that inhabit the cloudy surface of Jupiter are not as long-lived as the GRS. Three white ovals (Fig. 9-33) in the STB have been observed since 1939, making them the second oldest features on the planet. Named BC, DE, and FA, they are similar to the GRS in that they spin counterclockwise and are considered high-pressure centers. Since 1980, the white ovals have been growing smaller with an accompanying increase in their brightness. It is estimated that they are in the last few years of their existence, so a careful eye should be kept on their sizes, intensities, and transits.

In 1983, Jose Olivarez, ALPO Jupiter Recorder, discovered 12 features along the south edge of the NEB. These are blue spots with accompanying festoons that seem to be holes in the Jovian atmosphere. Because they are hard to discern against the disk, a red filter can bring them out for the observer.

In addition to the spots and ovals on the planet, the belt systems are home to a wide variety of activity. Primary among this activity are *disturbances* in the STrZ and the SEB. At the beginning of this century, from 1901 to 1939, the STrZ was home to a major disturbance that spread halfway across the planet. From 1939 until the mid-1970s, the dark feature seemed to disappear only to become quite prominent again in 1978. The disturbance seems to show a definite trend toward periodical flare-ups, so the area should be kept under observation. In 1985 a white spot was detected in the SEB that has grown to become a great white streak across the belt. The SEB bright streak disturbance has been seen to reestablish itself during each of the subsequent oppositions.

Filters are a great asset for the observer studying the various features on Jupiter. The colors on the planet range from the brownish belts to the bright, off-white zones. The contrast between belts and zones can be increased with a light blue Wratten 80A filter. Festoons and columns exhibit a bluish cast that can be darkened by using a yellow filter.

Drawing Jupiter

It's time to get out the pencil and paper again. Drawing Jupiter is different from drawing Mars. The primary drawback with this planet is its rapid rotation. Because of this factor, drawings must be executed rather quickly, in no more than 10 minutes. After this length of time the features on the disk will have noticeably shifted position, and the accuracy of the drawing will be destroyed. Drawings of Jupiter fall into two types: full-disk drawings, like those done for Mars, show the entire planet visible at the time of the drawing; strip sketches are used to portray details in a small band of latitude on the planet. Both types of drawings make important contributions to our growing knowledge of the giant planet, so let's begin with the full-disk technique.

As with Mars, the first thing to do is prepare a blank. Because Jupiter rotates so fast, it is not a perfect circle. The equator bulges out and is seen as being wider than the polar areas. Disks are a bit difficult to prepare. I would recommend that you use those supplied by ALPO because they have the properly drawn disk (Fig. 9-34). Because of the planet's rapid west to east rotation, seen right to left in an astronomical telescope, it is best for the observer using a refractor or SCT to leave out the star diagonal. This way you will not become confused by the image reversal given by the diagonal.

Begin your drawing (Fig. 9-35) by determining the position of the equator. Quickly fill in the main belts and zones, then add the fainter belts. Before you begin to fill in the detail, note the time and mark it on your form. You will have 10 minutes to place the main details on the belts and zones (Fig. 9-36). You must be constantly aware of the rotation of the planet and keep checking the position of each detail with relation to the rest of the drawing. Indicate bright, but ill-defined areas and low-contrast features with a dashed line. Use a sharpened eraser to add details to the darker belts. When all the details are drawn, and before your time is up, check over the details on the planet and compare the drawing to the image. Shade in the belts and prominent features to representative levels (Fig. 9-37) then note the time the drawing was completed. Because of the time limitation imposed by Jupiter's rotation, it takes a bit more practice to get the hang of making full-disk drawings. The only way to become proficient is by constant practice, so don't be disappointed by your early drawings and by all means save them, no matter how bad you think they are!

Strip sketches are a bit easier to execute because instead of having to range over the entire visible ball of the planet, the details sort of come to you. A strip sketch is simply a continuous record of the details in a particular zone or belt of the planet as it crosses the CM (Fig. 9-38). Time is noted for the transit of each feature in order to compute its longitude, which is labeled across the bottom of the sketch. Strip sketches allow the observer to concentrate on detecting and recording much finer detail than can be seen during a full-disk drawing. Like full disk drawings, strip sketches concentrate on obtaining the position and structure of the object, not on producing a work of art.

The technique for making a strip sketch is simple. If you are studying a single zone or belt, the boundaries of the belt are the boundaries of the illustration. If you are studying a complex area, like the SEB with its north (SEBn) and south (SEBs) aspects, indicate which is which by a notation at the edge of its placement on the page. Now sit back and watch the show. As each feature crosses the CM sketch away, moving across the page. Pay close attention so you can obtain as much detail as possible.

Filter_____

DATE (UT)_____

TIME (UT)_____

TELESCOPE _____

MAGNIFICATION_____

SEEING _____ TRANS. _____

OBSERVER _____

ADDRESS _____
_____

SYSTEM I _____ II_____

NOTES

Filter_____

DATE (UT)_____

TIME (UT)_____

TELESCOPE _____

MAGNIFICATION _____

SEEING _____ TRANS. _____

OBSERVER _____

ADDRESS _____
_____

SYSTEM I _____ II _____

NOTES

CENTRAL MERIDIAN TRANSITS:

TIME	SYS	LONG	DESCRIPTION

Fig. 9-34. ALPO Jupiter observation form.

Note the time of transit for the center of the feature; if you are totally engrossed either the preceding or following edge will do. As the feature under observation moves off the CM, another might be ready to take its place. If it appears that there will be a gap before the next feature reaches the CM, position it on the page a proportional distance from the last feature entered. It's not a bad idea to indicate the transit time of the point midway between the two features to ensure accuracy.

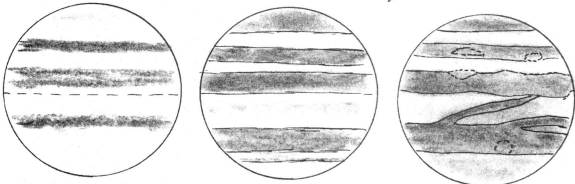

Fig. 9-35. Drawing Jupiter I.   Fig. 9-36. Drawing Jupiter II.   Fig. 9-37. Drawing Jupiter III.

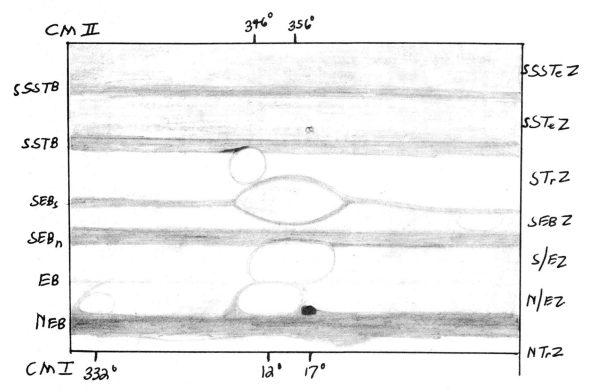

Fig. 9-38. A strip sketch of Jupiter. (Courtesy Dan Troiani.)

The beauty of a strip sketch is that you can extend it over as long a period of time as you wish with no time limit for its completion.

The Satellites of Jupiter

The show Jupiter puts on for the amateur astronomer is not restricted to its cloudy surface. There is much to see and study in the space around the planet. When Galileo first turned his telescope on the planet, he was surprised to find four small stars close to it. These four satellites, called the Galilean satellites (Fig. 9-39), are the only ones visible out of a total of 16 that circle the planet. They are also the subject of some basic scientific observations that can be made with the smallest telescope. In order out from the planet, they are : I—Io, II—Europa, III—Ganymede, and IV—Callisto.

During the course of their orbital journeys around the planet, each satellite can be subject to one of four possible events (Fig. 9-40). They can be eclipsed by the planet,

Fig. 9-39. Jupiter with satellite shadow passing across its face. (Courtesy NASA.)

they can be occulted by the planet, and either the moons themselves or their shadows can move across the planet in transit. Each of these events can further be broken down as entry to or exit from the event. These events can be studied with almost any size telescope.

The easiest of the phenomena to observe are the eclipses of the satellites. Since their discovery by Galileo, these four satellites and their movements into and out of eclipses have given us a chance to study the effects of the gravity of one body on another, called *perturbations*, in a closed, rapidly moving system. Observations of these events were also of tremendous value during the seventeenth and eighteenth century in refining the longitudes of positions on Earth. Today we have more sophisticated ways to determine longitudes on Earth, but the Galilean satellites and their eclipses still provide us with the best real-time laboratory for determining effects of perturbations.

Since 1975, ALPO has had a program for the timing and analysis of Galilean eclipses. Since the program's beginning, over 2,500 timings have been analyzed to refine the positions of the four Galileans. Not only is this information of use to theoretical scientists in determining perturbation effects, it is also of use to NASA in its planning and launching of the forthcoming *Galileo* mission to Jupiter.

The observational requirements for the program are simple. A telescope as small as 50mm, using at least 60×, and an accurate timepiece are all that are necessary to join in another scientifically useful program. Electronic watches manufactured today can give

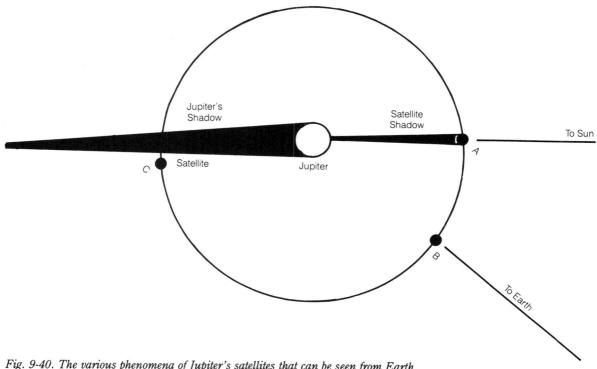

Fig. 9-40. The various phenomena of Jupiter's satellites that can be seen from Earth.

you accuracy to within one second when they are checked against radio time signals from station WWV. Each year ALPO's *Solar System Ephemeris*, the U.S. Naval Observatory's *Astronomical Almanac*, and the RCAS's *Observers Handbook* publish ephemerides for the events of the Galileans for the upcoming year. Your job is to observe as many of the eclipse events as possible during the next apparition, record the time of the event using your timepiece, and then forward the raw data to ALPO for analysis.

In the field, set up your equipment well in advance of the predicted event, which is listed for UT so don't forget to make the necessary corrections. You should be ready to go at least 10 minutes before events predicted for Io and Europa, and at least 30 minutes before those for Ganymede and Callisto. Use the extra time to identify the Galilean satellites present around the planet.

Eclipse events that occur before Jupiter reaches opposition cause the satellite to disappear long before it reaches the preceding limb of the planet. For Io and some points in the orbit of Europa, this means that you will probably not be able to see the satellite emerge from the shadow as that event will be blocked by Jupiter itself. The other satellites are far enough away that they will reappear before they pass behind the planet in occultation.

After opposition the reverse is true, with the satellites reappearing well away from the following limb of Jupiter. Disappearances after opposition take place in a small area defined on one side by the limb of the planet and on the other by the shadow of Jupiter. If you are timing a *disappearance*, locate the target moon and watch it fade slowly from view, recording the time to the nearest second when the last speck is visible. When recording the *reappearance* of a satellite from the shadow of Jupiter, take a guess where the satellite will appear and note the time to the nearest second when the first speck of light is visible. Transfer your data to the report form supplied by ALPO (Fig. 9-41) and fill in the necessary data on your telescope, seeing conditions, and any information you feel is relevant to your timing. When the sheet is full, mail it to ALPO and ask for another. They will be only too happy to send it along.

Transits of the Galileans and their shadows across the disk of Jupiter are also easy to follow, although the instrument used should be a bit larger. While a 60mm is sufficient to show the entry and exit of the satellite as a bright point of light, a larger instrument will give better resolution when following the satellite completely across the disk. Because the ingress and egress of the satellite requires you to resolve its actual disk, an aperture of at least 8 inches is required. The shadow of the satellite will show up as a dark, distinct black spot on the surface of the planet. It is easy to distinguish against the background of belts and zones because it is so black. The satellite itself is more difficult to detect, especially when it is crossing a zone, because it is a light object that can easily blend into the background of the planet.

As with eclipses, the sequence of transit phenomena is dictated by the location of Jupiter in its orbit. Prior to opposition the shadow precedes the satellite, and its movement onto the disk, or *ingress*, will come before that of the satellite itself. After opposition, the opposite is true with the shadow following the satellite so that it can still be seen on the planet after the satellite itself has left the disk, or *egressed*. Timing these events is simple, but requires a sharp eye. The crucial points during ingress are when the satellite, or shadow, first touches the disk (first contact) and when the satellite or shadow is seen to be completely on the disk (second contact). Egress is harder to detect

OBSERVER NAME Michael R. Porcellino

APPARITION [YEAR(S)] ___1988___

Describe Time Source and Estimated Accuracy:
Electronic watch set to radio time signals (WWV)/ accurate to 1second

Event Type (a)	Pred. UT Date (b)	Pred. UT Time (c)	Observed UT Time (d)	Type (e)	Aper. (cm.)	Magnif.	See-ing	Trans-parency	Sky Br.	Notes(g)
						Telescope Data		Sky Conditions (0-2 scale)(f)		
3 D	0102	03 33	03:38:45	R	8	100x	0	0	0	
3 R	0102	05:52	05:48:31	R	8	100	1	0	0	
2 D	0104	02:01	02:02:20	R	8	100	2	2	2	Clouds moved in at mid-event. Prob. affect timing
2 D	0111	04:41	04:42:40	R	8	100	2	0	1	estim close to Planet.
			satellite egress occult							at 0420 UT

NOTES: (a) 1 = Io, 2 = Europa, 3 = Ganymede, 4 = Callisto; D = Disappearance,
R = Reappearance.   (b) Month and Day.   (c) To 1 minute.
(d) To 1 sec., corrected for watch error if applicable. Indicate in Notes if
the observed UT date differs from the predicted UT date.
(e) R = Refractor, N = Newtonian Reflector, C = Cassegrain Reflector, X =
Compound/Catadioptric System; if other type, indicate in Notes.
(f) These conditions, including sky brightness (due to moonlight, twilight,
etc.) should be described as they apply to the actual field of view, rather
than general sky conditions. Use whole numbers only, as follows:
   0 = Condition not perceptible; no effect on timing accuracy.
   1 = Condition perceptible; possible minor effect on timing accuracy;
   2 = Condition serious; definite effect on timing accuracy.
(g) Note wind, drifting clouds, proximity of satellite to Jupiter, etc.

Fig. 9-41. ALPO Galilean Satellite Eclipse Timing Report Form.

because the satellite may blend into the background. Timing should be made when the satellite or shadow reaches the edge of the disk (third contact), and when it is completely clear of the disk (fourth contact). Again, timings should be made to the nearest second.

Occultations of the Galileans are timed in much the same way as those of a star occulted by the Moon. Time is recorded when the satellite disappears behind the planet and then when it reappears on the other side of the planet. The reappearances of the satellites are particularly hard to catch because of the glare from the planet itself.

OBSERVING SATURN

The last planet in the amateur astronomer's Solar System is far and away the loveliest. Saturn, with its ring system, is probably the most sought after sight by people attending public star parties. It is easy to tell when the person at the telescope encounters the planet, whether for the first or the umpteenth time. They simply freeze at the eyepiece, not daring to move for fear of losing the wonderful sight. On many occasions they have asked me if they were really looking at Saturn or a cleverly placed picture of the planet.

Observing Saturn pushes the amateur to the very limit of his equipment and abilities. Despite its beauty, Saturn is a very difficult object to observe.

When Galileo first turned his telescope on the planet, he was puzzled by its appearance, recording it as a "triple planet". His instrument could not resolve the majestic rings of the planet into anything more than "a large disk, touched on either side by a smaller disk." It would remain for Christian Huygens to solve the riddle of the triple planet some 50 years later.

During its apparitions Saturn is easily visible for 10 months out of the year, varying in magnitude from +1.5 to 0. To the naked eye, Saturn is easy to pick out from the background stars as a bright yellow-white star. The apparent diameter of the disk of Saturn varies from a minimum of about 15 arcseconds to a maximum of 21 arcseconds.

Saturn, like Jupiter, is a gas giant and it shows the characteristic dark belts and lighter zones. But unlike Jupiter, you are going to need a larger instrument to see them. A telescope of at least 4 inches in aperture and preferably 6 inches or larger is needed to pick out the subtly colored bands from the bright background clutter. Saturn also does not display the abundant atmospheric activity of its inner neighbor. The most prominent features on the Saturnian disk are white spots and ovals similar to those on Jupiter. But unlike those on Jupiter, these are relatively short-lived, lasting only a few weeks.

Saturn is one planet that almost requires the use of filters during observation. The lack of contrast between the features, when they are there, and the ball of the planet make it difficult to pick them out of the background. This is further complicated by the bright image transmitted by the ball of the planet itself. Because the belts on the ball are rather dusky, they are often lost in the glare of the planet. A light blue filter (Wratten # 80A) will give you your best chance at separating the belts and zones by increasing their contrast.

ALPO's Saturn Section distributes a complete kit of materials, including planetary blanks for all the aspects of the rings, and information on the planet's elusive features. If you want to challenge your observing skills, join ALPO and get into the Saturn Section.

The Rings of Saturn

The whole show for the small telescope pointed at Saturn is the rings. Less than 100 feet thick, they measure 175,000 miles across and comprise a celestial hail storm whipping around the planet at 90,000 miles per hour. The brightness of the planet, as we see it with the naked eye, is due in large part to the reflective properties of the rings. Saturn takes 29.5 years to circle the Sun. Twice during that time, once after a 13.75-year period and again after a 15.75-year period, the Earth passes briefly through the plane of the rings. For this period of time Saturn loses its most prominent feature, and the ball of the planet is bare in the telescope. Prior to this event the rings appear to slowly close and there is a resulting loss of brightness when seen with the naked eye. The same orbital mechanics that briefly take the rings away can show them to us in all their glory. At a point between each plane passage the planet is tipped toward Earth by 27 degrees, displaying for us almost the entire ring system. During the 13.75-year period the ring system's southern face is presented to Earth, while during the 15.75-year period the northern face of the rings are seen.

If you look at the planet before opposition, you can detect the shadow of the planet as a jet black notch on the western side of the rings. After opposition the notch moves around to the east side. Associated with the shadow is a bright white spot, called the Terby White Spot. This spot is thought to be caused by electrical activity brought on by the shadow's passage over the rings.

The rings of Saturn are not a single feature, but a system of rings and gaps that circle the planet. From Earth the amateur can easily detect the three main rings, labeled ''A,'' ''B,'' and ''C'' (Fig. 9-42). The outermost ring, ''A,'' is itself divided into two sections, the outer one dark and the inner one a bright *annulus* separated by a fine, dark line called the *Encke's division*. Theoretically a difficult object for an 8-inch telescope, Encke's division has been reported by numerous observers using 6-inch instruments. These reports are confirmations of the dictum stated by Sir John Herschel, William's son, that ''when an object is discovered by a superior power, an inferior one will suffice to recover it.'' Normally seen extending through the ring a short distance, Encke's division seems to show no definite location along the division of the ring.

Separating rings ''A'' and ''B'' is a prominent dark division called *Cassini's division*. The brighter inner surface of ring ''A'' as seen by a 6-inch telescope exhibits an irregular appearance along the edge fronting Cassini's division. Cassini's division itself is easily seen in a small telescope as a dark gray gap between the two rings.

Ring ''B'' is the brightest and largest of the three major rings. Its appearance in a telescope shows its brightest at its outer edge, then darkening as it nears the planet. As many as seven dark lines can be seen in Ring ''B'' with telescopes as small as 5 inches in aperture.

The innermost ring, Ring ''C,'' is also called the *Crepe ring*. Having little contrast against the dark background of the space between the planet and the rings, it is most easily seen as it crosses the disk against Saturn's bright background.

The Satellites of Saturn

Saturn is blessed with 17 satellites, 5 of which can be seen with a small telescope. The brightest of these, Titan, is the largest satellite in the Solar System and appears as a

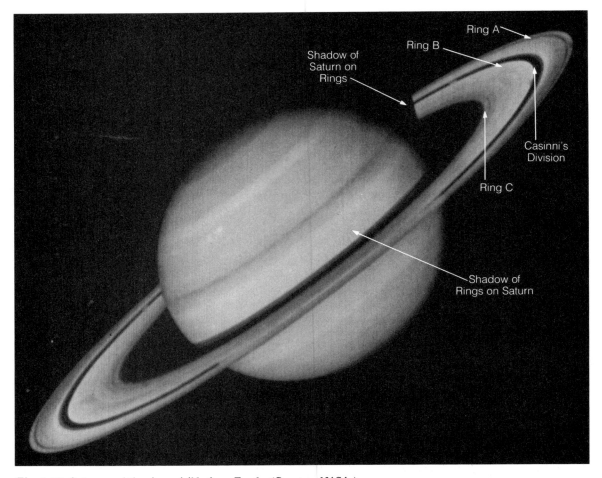

Fig. 9-42. Saturn and the rings visible from Earth. (Courtesy NASA.)

magnitude 8 star that orbits the planet in 16 days. Closer in is the moon Rhea, visible as a magnitude 10 star that orbits the planet in about 5 days. Inside the orbit of Rhea an amateur with a 4- to 6-inch telescope can locate Tethys and Dione. Tethys, orbiting the planet in 2 days, appears as a magnitude 10 star, as does Dione, which completes a circuit of Saturn every 3 days.

The outermost visible moon, Iapetus, requires almost 80 days to orbit the planet and has a variable magnitude ranging from 10.1 to 11.9. The variability of Iapetus is caused by differences on the surface of the satellite's two hemispheres. One is covered with highly reflective ice, while the other side has been stripped of this ice cover by some type of action.

10

Observing
Comets and Meteors

The heavens themselves blaze forth the death of princes.

Shakespeare

Most professional astronomers regard comets and meteors as annoying little smudges and streaks of light that intrude on their photographic plates. But there has always been a special bond between amateur astronomers and the "rubble" of the Solar System.

While professional astronomers tend to ignore these small members of the Sun's family, amateurs seek them out. An amateur will spend countless hours searching for a new comet, or an entire night counting streaks of light emanating from a meteor shower. Each of them offers a different reward, and each of them offers the amateur a way to contribute to science.

COMETS

Amateur astronomers and the public react differently to comets. To most people a Comet is a household cleaner. Few members of the nonastronomical public have seen a real live comet. They may have seen pictures of one, usually comet Halley in 1910, but pictures don't count. Professional astronomers come out of their observatories and proclaim the coming of very bright comets, like comet Kohoutek in 1973.

People respond with interest to these announcements because they want to see the spectacle presented by a really bright comet. On the appointed day they dutifully go outside and look toward the appropriate place in the sky and see nothing. They shrug their shoulders and go back inside to the light of the electronic fireplace. When a really spectacular comet shows up, like comet West in 1975, the people listen with interest but

191

remember that two years earlier these same people talked about another comet, the name of which escapes them. They do remember that they looked but didn't see anything, so they shrug their shoulders again and stay inside, thus missing a true celestial event.

Amateurs don't react that way. They are out there looking at the comet, no matter how faint. They look because there is no one else out there to look and because each comet gives us a chance to peek at the history of the Solar System. Professionals may want to turn their instruments on the comet as it streaks toward the Sun, but the time on their telescopes is too expensive to divert them to watch such a thing. Exceptions do exist, the return of comet Halley in 1986 being a case in point. Then everyone was interested—public, amateur, and professional alike. But when the United States decided not to send a spaceprobe to the comet, interest in the professional community waned slightly. When the comet did not live up to its spectacular buildup, interest dropped like a rock in the public sector (Figs. 10-1 and 10-2). The only ones who really saw the comet were the amateurs.

Cometary Geography

We never see the heart of the comet because it is too small and too far away. It lies in the cold of the *oort cloud*, a huge shell of material leftover from the formation of the Solar System. It is a cold, lumpy mass of ice and dirt orbiting a Sun, which appears like another of the countless stars in the sky. But then something happens, such as a faint tug or a push by a more massive object, and the heart of the comet finds its orbit changed ever so slightly. The long journey to the Sun has begun.

On its way to the nearer reaches of the Solar System, the heart or nucleus of the comet begins to change. The driving rain of radiation from the rapidly brightening Sun causes it to heat up. Gases begin to bubble and boil off. A cloud of gas called a *coma* surrounds the nucleus and reflects the light of the still distant Sun making the comet visible against the darkness of space. From Earth its light may be detected on photographic plates made by an observatory, or it may appear as a faint smudge in the field of an amateur astronomer's telescope. Regardless of who sees it first, a new comet is discovered and given a name.

During the course of their brief and rapid race around the Sun, comets carry three names. The first is the designation of the year of discovery and the order of its appearance after other comets discovered that year. The first comet discovered in 1988 would carry the designation 1988a, the second 1988b, and so on. The comet's second name is the name of its discoverer. The tenth comet discovered in 1988 was found by California amateur Don Machholz. This comet now carries the names 1988j and Comet Machholz. When it reaches its closest point to the Sun, or perihelion, Comet Machholz will receive its third name, the year it reaches perihelion followed by a roman numeral indicating its order among all the comets reaching perihelion that year. So if Comet Machholz is the last of 15 comets reaching perihelion in 1988, it will carry the additional designation 1988XV.

Changes have been taking place on the comet. The coma and the nucleus it hides are now called the *head* (Fig. 10-3). It is reacting more and more to incoming radiation

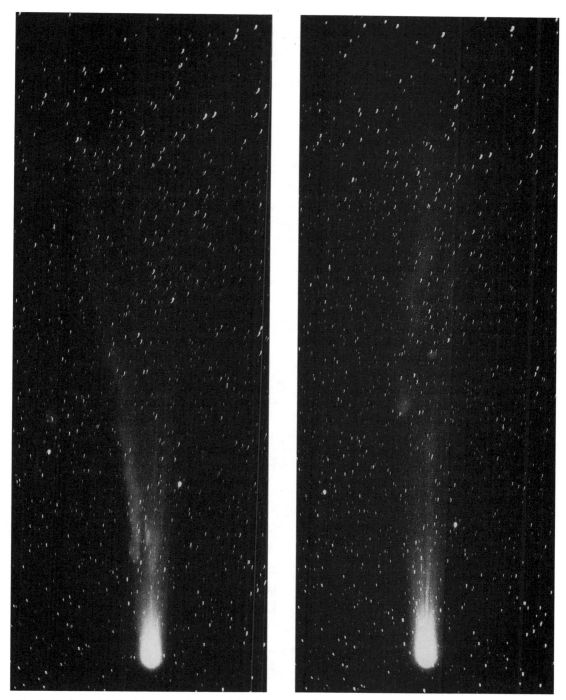

Fig. 10-1. Comet Halley in 1910. (Courtesy Lick Observatory.)

from the Sun. This causes more of the nucleus to boil off, carrying the dust and dirt of the nucleus away from the head. This dust is caught by the *solar wind*, a wave of highly energized particles thrown out from the Sun, and is pushed away from the head of the comet where it begins to glow from the radiation. We see this glowing dust as the *dust tail* (Fig. 10-4). The dust tail of the comet points along the direction of the solar wind, away from the Sun.

A comet really has more than one tail. In addition to the dust tail, a gas or *ion tail* can be seen streaming out directly behind the comet's head. Sometimes another tail, an *anti-tail*, is seen pointing toward the Sun (Fig. 10-5). The anti-tail is nothing more than an optical illusion created when we look through the plane of the comet's orbit at the dust tail.

As the comet sweeps around the Sun, it may or may not reach the safety of the Oort Cloud again. Minute tugs of gravity from each body it passes during its journey toward the Sun act to change its orbit around the Sun. The orbit of a comet determines the ultimate fate of the comet (Fig. 10-6). If the orbit is an elongated circle called an *ellipse*, it will swing around the Sun and head back toward the Oort Cloud. It may not make it that far, however, because its speed has been affected by the huge gravitational pull of the Sun. As it heads away from the Sun it begins to slow down. At some point in its orbit it will stop and begin to fall back toward the Sun. A comet with this type of orbit will return to the Sun again and again until it ultimately melts away to almost nothing.

These comets are called *periodical comets* because they return for definite periods. Comet Halley is a periodic comet. Its orbit takes it out to the orbit of Neptune over a

Fig. 10-2. Comet Halley in 1986. This return of Halley was a disappointment to the public. (Courtesy Lick Observatory.)

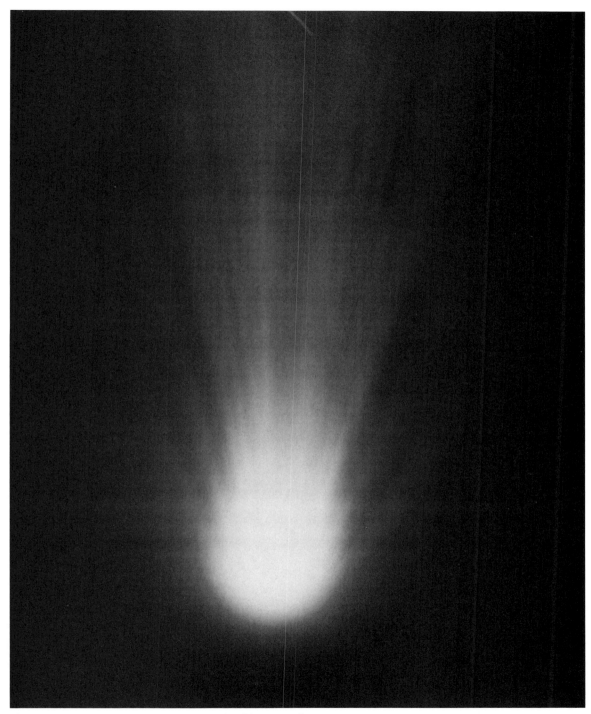

Fig. 10-3. The head of a comet. (Courtesy Lick Observatory.)

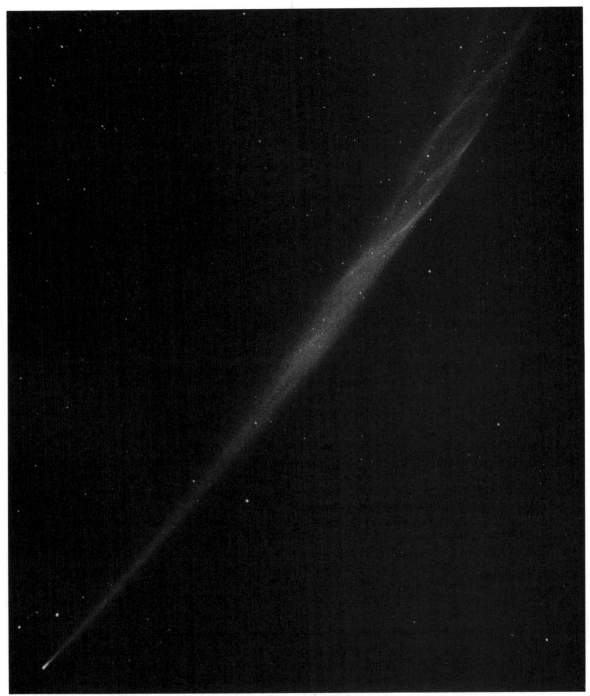

Fig. 10-4. Comet tails come in all shapes and sizes: (A) comet Ikeya-Seki (1965f). (Courtesy Lick Observatory.)

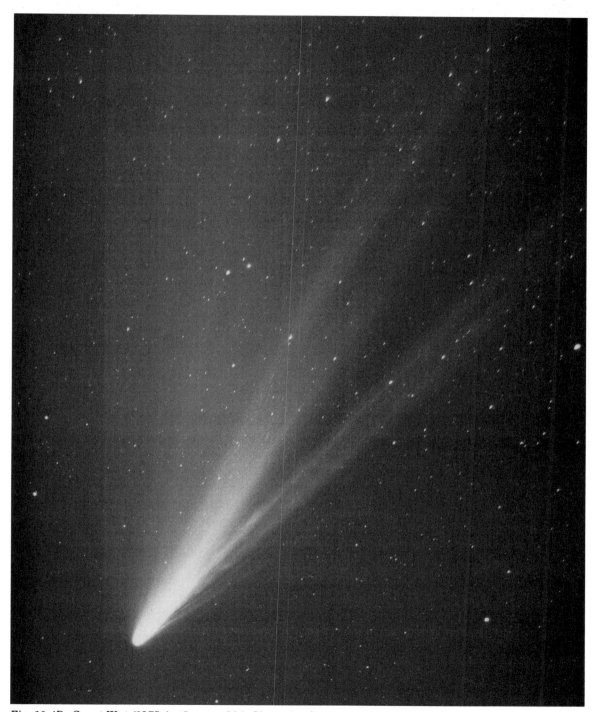

Fig. 10-4B. Comet West (1975n). (Courtesy Lick Observatory.)

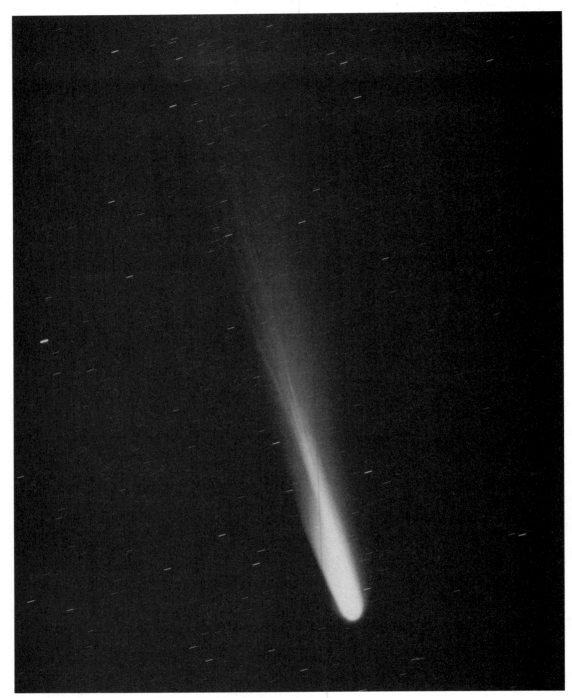

Fig. 10-4C. Comet Bennett (1969i). (Courtesy Lick Observatory.)

Fig. 10-5. Comet Arend-Roland with anti-tail. (Courtesy Lick Observatory.)

period of 76 years. Comet Encke is another periodic comet, but Encke doesn't venture so far from the Sun. It only goes out as far as the orbit of Jupiter and has a period of 3.3. years.

Periodic comets fall into two classes. *Short-period comets*, like Halley and Encke, have periods of 200 years or less. Comets that have periods over 200 years are called *long-period comets*. The names of all periodic comets are preceded by the designation "P/," so comet Halley becomes P/Halley.

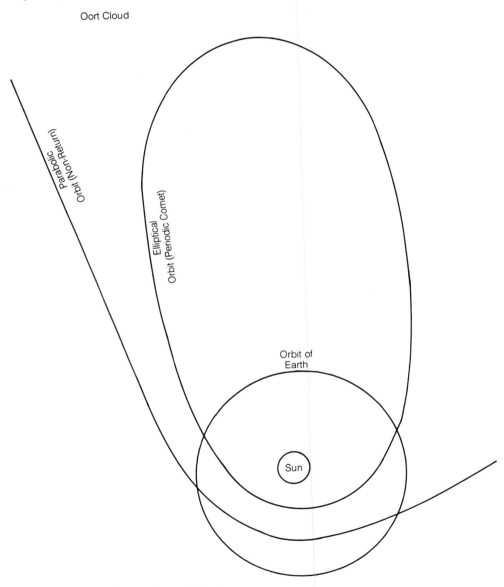

Fig. 10-6. *Comet orbits and the oort cloud.*

The Role of the Amateur

Amateurs have always had a special relationship with comets. Not only do they provide valuable information on their structure, but amateurs very often make the ultimate contribution, the discovery of comets. Comet hunting has always been an area in which amateurs have excelled. Many comets are discovered by accident when they turn up as smudges on photographic plates made by professional observatories. Satellites like the infrared astronomical satellite (IRAS) have even been the first to detect comets on their way toward the Sun, but these discoveries also have been accidental. Amateurs throughout the world, however, are at their telescopes every clear night with the purpose of discovering comets. David Levy of Tucson, Arizona, has logged thousands of hours at his telescopes hunting comets in the richness of the sky. His efforts have been rewarded with four comets that carry his name.

Comet hunting is not for every amateur. If you want to hunt comets to gain fame and fortune, you are going to be disappointed. The hours are long and there is no guarantee that you will ever find a comet. Don Machholz searched for 1,700 hours before he found his first comet. David Levy spent over 900 hours at his telescopes before he bagged his first comet. These numbers, when combined with such mundane things as working a regular job and having a social life, tend to discourage all but the most dedicated observers.

If you insist on going out tonight and looking for a comet that will carry your name for all eternity, there are a few things you can do to help the process along. Comet hunting is a competition between amateurs and professionals. While most professional astronomers discover comets during the pursuit of other projects, a few have formed organized patrols and regularly use their larger, more sensitive photographic instruments in an attempt to recover periodic comets headed back toward the Sun. They are looking to confirm the computations made for the orbits of these comets. If they discover a new comet in the process, it's ''icing on the cake,'' so to speak.

Most of these professional efforts are centered on areas surrounding the meridian where the photographs can record objects as faint as magnitude 20. It would seem obvious that if you search this area, you are searching against some heavy opposition. The most fertile area of the sky, however, is the area that professional photographic patrols have to avoid. Recording objects within 45 degrees of either side of the Sun reduces the effectiveness of these professional instruments. Exposures made with the Sun just below the horizon can cut the limiting magnitude of the photograph and possibly damage the expensive, sensitive instruments. So the area near the Sun, just after sunset or before sunrise, is best for visual searching.

The tools of the comet hunter may include any sort of optical aid. You can even use the naked eye to search for bright comets near the Sun. William Bradfield of Australia, the reigning champion with 13 discoveries to his credit, has searched for comets for more than 2,300 hours over the last 17 years. During that time he has used everything from 7×35 binoculars, with which he discovered a comet in 1980, to a 10-inch reflector, with which he discovered a comet in 1984. Leslie Peltier discovered 12 comets using a 6-inch, f/8 refractor as his prime instrument. David Levy uses a 16-inch f/5 reflector as his primary instrument. In comet hunting the altazimuth mounting is a definite advantage over the equatorial. Because you will be observing near the horizon, the altazimuth allows you to easily scan along it without putting a strain on your back or your patience.

Focal ratio, while not crucial is important. It must be remembered that you are going to be looking for a very diffuse, nebulous object. While the images provided by f/4 and f/10 systems will look equally bright, provided the magnification is the same, the f/4 will let you see a wider area of the sky in your field of view and make the comet stand out better from the background. The f/10 covers a smaller area and, because it provides an appreciably darker field with a corresponding decrease in the brightness of a cometary object, it makes a suspected comet blend in better to the background.

It is best to use an eyepiece that gives a wide, flat field and a magnification of no more than 2D. Erfles are hard to beat in this category, although new designs like the Tele-Vue wide-field and the Nagler-2 are also useful. If you are using a short focal length newtonian such as a 6-inch f/4, you will want an eyepiece with a focal length between 40mm and 48mm. You can use any telescope or pair of binoculars to find a comet. The important thing is how you use the instrument.

When and Where to Look

The best time to look for comets is either just after sunset or before sunrise. Of the two, the time before sunrise yields the most new comets. A problem you will have to contend

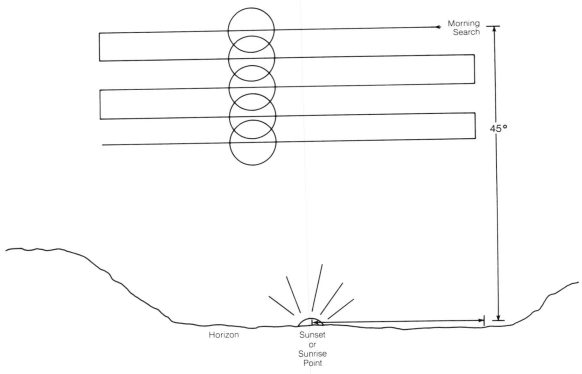

Fig. 10-7. Search pattern for comets. For a morning search, start about three hours before sunrise and follow pattern. For an evening search, begin about one hour after sunset and start at the horizon working up to 45 degrees. The fields of view for each sweep should overlap.

with is the Moon. When the Moon is beginning a new lunation and is rising brighter each night, it is time to search the sky before sunrise. As the Moon starts to spill over to the morning sky, try to time your searches to begin with moonset.

For the period just before a new Moon, when it rises about two hours before the Sun, begin your search about three hours before sunrise. This will give you at least one hour to search. If you are searching the sky after sunset, the best time to look begins about two days after a full Moon. This gives you at least an hour and a half from sunset until moonrise to hunt. As the Moon rises progressively later each night, you have more time each night for your search.

The area of your search should extend from the horizon to a point about 45 degrees above the horizon and within 45 degrees on either side of the point where the Moon has set or will rise. Cover the area with overlapping sweeps, each sweep arc containing slightly less than the field covered by your telescope (Fig. 10-7). For searches after sunset begin at the horizon and work up. Morning sweeps should begin at a point about 45 degrees above the horizon and work down.

What you are looking for is a little fuzzy patch of light (Fig. 10-8). During the course of your sweeping you will come across dozens, even hundreds of objects that fit that description. The skies are filled with ''little fuzzy patches,'' seemingly put there to confuse and confound the comet hunter. Charles Messier, the French comet hunter of the late eighteenth century, became so frustrated with these little fuzzies that he began to list them ''so that astronomers would not confuse these nebulae with comets just beginning to shine.'' Today we know this list as the *Messier Catalogue*. It includes 110 of the finest deep-sky objects in the heavens. In order to better prepare yourself and learn what comets do not look like, check out the members of this list before you start hunting.

In addition to the 110 objects on Messier's list, there are another 8,000 galaxies, nebulae, and star clusters listed in the *New General Catalogue of Nonstellar Astronomical Objects*, called the NGC for short. In addition to the NGC there is the *Uppsala General Catalogue of Galaxies (UGC), the Catalogue of Selected Non-UGC Galaxies* (UGCA), the *Catalogue of Galactic Planetary Nebulae* (PK), etc. I think you see what I'm driving at.

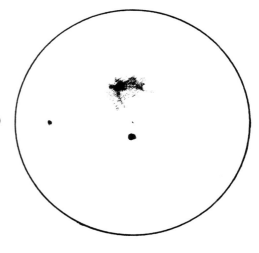

Fig. 10-8. Sketch of a comet as it appears in an amateur telescope. (Courtesy Dan Troiani.)

There are an awful lot of objects out there that you might confuse with a comet. Some areas are so choked with faint galaxies, like Coma Bernices, Leo, and Virgo, that it doesn't pay for the comet hunter to search there. So what do you do to sort everything out?

Get yourself a good star atlas. An atlas like Tirion's *Sky Atlas 2000* is an indispensable asset for the comet hunter. Not only does it list stars down to magnitude 8, it also lists thousands of NGC, and other objects. *Uranometria* lists stars down to magnitude 9.5 and thousands more deep-sky objects than *Sky Atlas*. A good atlas gives the comet hunter a reference to check quickly if he finds something suspicious. If it's listed, it's on to the next field. If it's not....

Reporting Your Find

Stay calm and check the image of your "comet" carefully. Is the object real or a light reflecting off a nearby star? Move the object around the field and see what happens. Check the object with a higher-power eyepiece. Is it a faint star near the limiting magnitude of your telescope or a group of stars? Both types of objects can look diffuse and nebulous. Draw a chart showing the object and the surrounding background stars. Be careful to get the distance relationships between the stars and your object correct. You can then transfer this location onto your star atlas and get an accurate right ascension and declination from it.

Comets move very quickly. By checking the drawn field with the field of view just 15 minutes later, you might be able to detect movement. If you are not sure, wait. Movement and the direction of that movement will be important when you finally report your find. So make sure it is moving and that you plot the direction of that movement on your chart. Does it have a tail? The tail may be faint, but if it is there and there is movement, it's usually the clincher.

Estimate the magnitude of your comet by comparing it with other stars in the field, which you have already identified from your atlas and chart. The best way to do this is to memorize how the comet looks and throw the field slightly out of focus. Compare the image of the focused comet to the now diffuse star images and decide which one it looks most like.

You're almost ready to take the next step, but don't count your comets before you're absolutely sure. Yes, it might be a comet, but it might not be yours. It could be a previously listed comet or a periodic on its way around the Sun. Check sources like the *International Comet Quarterly*, which lists orbital elements and ephemerides for all periodic comets due during the year. *Harvard Announcement Cards*, to which you can subscribe, announce many astronomical discoveries other than comets and send them to subscribers complete with ephemerides.

If the object moving, with or without a tail, is not a listed or misidentified deep-sky object, and is nobody else's comet, it's all yours. High-tail it into the house and call Western Union. It's time to send a telegram. The International Astronomical Union (IAU) maintains a clearinghouse for all comet discoveries at the Smithsonian Astrophysical Observatory in Cambridge, Massachusetts, called the Central Bureau for Astronomical Telegrams (CBAT). The IAU has developed a code to keep submitted information neat and uniform, although you can send the data in plain English if you are not sure of your coding abilities. The telegram is composed of a series of words and numbers. Let's say your name is Smith and you have discovered a comet near the star Beta in the constellation Hercules. Here is what your telegram to CBAT would look like.

SMITH COMET SMITH
20001 81031 08000
16200 22025 01094
78527 39315
MARY SMITH, 123 STREET
CITY, STATE

The first word is the name of the discoverer (Smith). The second word is what she has discovered (comet). The third word is the name of the observer (Smith).

The first group of numbers (20001) is the epoch of the coordinates that follow (2000). The *1* that follows 2000 means the position is approximate and probably plotted on an atlas like *Sky Atlas 2000*. A *2* would mean that the position is exact.

The next group of numbers (81031) gives the date of the observation. The first number *8*, is the last digit of the year 1988; *10* indicates the month of October; and *31* indicates the 31st day in UT.

The third number group (08000) indicates the time of the discovery in UT in decimals of the day. Each hour is .04 of a 24-hour day, so 08000 would indicate the observation was made at 02 hours 00 minutes.

The fourth series (16200) is the right ascension of the object in hours, minutes, and tenths of a minute. The object was seen at RA 16 hours 20.0 minutes.

The next group (22025) is the declination of the object. The first number, *2*, means that the declination is positive or north. If it was south the first number would be *1*. The declination of the object is +20 degrees 25 minutes.

The sixth group of numbers (01094) indicates the object's magnitude. The first number in this sequence is always *0*; the second number *1*, indicates that the magnitude is for the entire comet. If the magnitude was for the head alone this number would be *2*. The next two numbers, *09*, indicate that it is magnitude 9 and the final number *4* indicates the comet's appearance. CBAT has a scale listing information on the comet's appearance.

0—no info on head, no info on tail
1—no info on head, no tail seen
2—no info on head, tail less than one degree
3—no info on head, tail longer than one degree
4—head diffuse, no tail
5—head diffuse, tail less than one degree
6—head diffuse, tail longer than one degree
7—head diffuse with central condensation, no tail
8—head diffuse with central condensation, tail less than one degree
9—head diffuse with central condensation, tail longer than one degree

The next group (78527) is called a check sum. You get this number by taking the last five digits of the sum of all the number groups in the report.

The final group of numbers is another check sum. This time use the number groups that provide information on the object's position and magnitude, groups 4, 5, and 6. Again, use the last five digits of this sum.

The last information on the telegram is the full name and address of the observer.

Tracking a Comet

Once a comet is discovered, what can the amateur do to continue its study? The most important role the amateur can play is to keep a continuous watch on the comet with an eye toward its development and changes. Detailed drawings of a comet can show the growth or decline of the tail and the corresponding changes in the head of the comet. The Comet section of ALPO recommends that observers watch for the following details.

The size of the coma should be estimated or measured during each observation session. This can be done by comparing the coma size to the distance between two stars. You can also use a micrometer reticle eyepiece (see chapter 11), which allows you to align the coma across a graduated line for a more precise measurement. You can measure the changes in brightness by estimating the coma's *degree of condensation* (DC) on a scale from 0 to 9, with 0 being a diffuse, uniformly bright coma and 9 being a sharply defined, starlike coma.

Make an estimate of the length of the tail in degrees and tenths of degrees. To obtain this information, accurately plot the comet on a star chart and then measure the desired dimensions directly from the chart. Also, try to determine the *position angle* (PA) of the tail. To do so, measure eastward along the coma from the north. Use 90 degrees as east, 180 degrees as south, and 270 degrees as west. If you can't figure out which way is north, give the telescope a nudge toward Polaris. The edge of the field through which stars enter is north. Check the tail to see if it is smooth or if it has developed any condensations or twists since your last look.

METEORS

What's that you say? You have no telescope or binoculars but want to make a contribution, any kind of contribution, to astronomy? Well don't fret, I've got just the thing for you. Meteor observing!

Meteors are those streaks of light that flash across the sky. They are small particles of dust and dirt that run into the Earth's atmosphere and burn up. Meteors can appear in the sky as random events called *sporadic meteors*, or they can come from a common point in the sky, called a *radiant*, and be associated with a *meteor shower*. Showers are usually the left over debris of a comet that has disintegrated after too many trips around the Sun. Remnants of the comet, in the form of dust and minute solid particles, remain in the orbit once occupied by the comet. If the Earth happens to travel through the plane of this orbit, we see a meteor shower.

The equipment needed for meteor observing is simple. A reclining lawn chair, a clipboard, paper, and pencil is all that is necessary. The procedure for observing is just as simple. Select an area of the sky, sit back, and make a note of each meteor that crosses the area you have under observation. Can't get much easier than that can it? Despite its apparent ease, meteor observing, especially during times of a shower, can provide useful information. The procedure is the same, but there are a few more things for which to look.

First, try to determine if the meteor is a member of a shower or if it is a sporadic. To do so hold your clipboard, or any straight edge, along the path of the meteor and note where it came from. Shower meteors will appear to originate from a radiant in the same general area of the sky. The shower is named for the constellation that contains that radiant. The Perseids, a show that occurs during August, all originate from a spot in the constellation Perseus. Table 10-1 contains a list of the major meteor showers that take place each year. Any meteor you trace back to that constellation using your straight edge will be a member of that shower. Mark any shower meteor with an abbreviation denoting the shower to which it belongs, such as PE for Perseid. Mark any sporadic meteors with the letter ''S.''

You should also estimate the magnitude of each meteor you see. As the meteor crosses the sky it will appear to grow brighter before it fades out. Estimate its magnitude at the brightest point of the meteor's path. This is simply a matter of comparing that brightness to stars in the area. If a meteor flashes near a third-magnitude star and appears just as bright mark it down as magnitude three.

Estimate the altitude above the horizon where the meteor was seen in order to determine the amount, if any, of atmospheric extinction discussed in chapter 7. If the third magnitude meteor was seen about two-thirds of the way between the horizon and the zenith, its altitude would be recorded as 600 degrees.

Meteors can vary greatly in brightness. Some meteors may be brighter than the planet Venus, magnitude −4. These are called *fireballs*. If a fireball explodes it is called a *bolide*. Fireballs, bolides, and regular meteors can all leave trails of smoke in the sky called *trains*. Note if the meteor leaves a train approximately how long the train stayed visible.

The record sheet you use should have room to record all this information. You should also note the beginning and ending time of the observing session a well as the date. Ruled paper allows you to run a column of numbers down the left side of the page while across the page you list the time of the event, in UT of course, whether the meteor was a member of a shower or a sporadic, its magnitude, altitude, and any remarks such as color or if it left a train.

Table 10-1. Major Meteor Showers

Shower	Dates	Max.	Rate
Quadrantids	Jan 1- 5	Jan 4	40
Lyrids	Apr 19-24	Apr 21	15
Eta Aquarids	May 1-8	May 5	20
Delta Aquarids	Jul 15-Aug 15	Jul 28	30
Perseids	Jul 25-Aug 18	Aug 12	50
Orionids	Oct 16-27	Oct 21	30
Leonids	Nov 15-19	Nov 17	75+
Geminids	Dec 7-15	Dec 14	60
Ursids	Dec 17-24	Dec 22	20

You can see more meteors after midnight when you face the direction toward which the Earth is traveling through space. Meteor observing can be a very late project so dress warmly to view the spring and autumn showers. The shower radiants may rise early in the evening, but take a few hours to reach their highest point in the sky. The higher the radiant the better the meteor count and visibility because the light from the meteors must travel through less atmosphere.

While meteor observing may not have the glamour of comet discovery, it is still basic science that needs to be done. Besides, doing science from a lounge chair is plain, simple fun!

11

Observing
Double Stars

*Many parts of the sky...are full of the most interesting and
beautiful groups and combinations.*

Rev. T. W. Webb

Since Giovanni Riccioli discovered the first one with a telescope in 1650, astronomers have been fascinated by double stars. In the 1700s Sir William Herschel began his investigations into stellar parallax by examining, measuring, and cataloging some 269 pairs of double stars. During the course of his work, he determined that some stars had a common orbital motion, thus discovering binary star systems. His pioneering work on the subject is considered the prime impetus for the study of double stars. Following in his father's footsteps, Sir John Herschel continued the study of double stars, reexamining his father's original list and adding 380 doubles of his own.

In 1814 an astronomer named F. G. Wilhelm Struve began work at the Dorpat Observatory in Estonia. Poorly equipped but with ambitious plans, Struve began a study of all the then known double stars. When his initial study was completed, he had compiled a list of 795 doubles, which he published as a catalog in 1822. Having gained favor and patrons with his impressive work, Struve was able to update and modernize his equipment. During a two-year period from 1825 to 1827, Struve put his new equipment to good use surveying 120,000 stars and discovering 3,112 new pairs. Continuing his measurements and study, he accounted for another 2,343 discoveries over the next decade.

In 1839 he became director of the newly established observatory at Poulkova. There he began his new project, the examination of all the stars in the Northern Hemisphere down to seventh magnitude. Struve was forced to turn the project over to his son Otto when the project was only a month old. Like the Herschels, the younger Struve carried on the family tradition and added another 514 new doubles by the time the project was completed in 1842.

Professional astronomers were not the only observers to study these beautiful objects. In England a retired admiral set up a private observatory and began his own study of doubles. William Smyth, working with good optics but poor measuring equipment, compiled a list of 680 pairs from 1830 to 1843. Also in England, the Reverend W. R. Dawes began work on doubles that would span 37 years. Called "eagle-eyed" by the Astronomer Royal Sir George Airy, Rev. Dawes published a catalog in 1867 that was received as a work of professional caliber.

It was no different on the continent. In 1852 Baron Ercole Dembowski began to make measurements with a 5-inch telescope outside of Naples. Over the next seven years, Dembowski measured over 2,000 double stars. Between 1862 and 1878, he made over 21,000 additional measurements and received the Gold Medal from the Royal Astronomical Society for his contributions to the study of double stars.

In 1870 a gentleman from the United States entered the world of astronomy and left an indelible mark. As Clerk of the U.S. Circuit Court in Chicago, S. W. Burnham found he had ample time on his hands. The purchase of a 3-inch refractor gave him a way to fill that time. Fascinated by his newfound avocation, Burnham soon found himself drawn into the demanding study of double stars. Purchasing a 6-inch, Alvan Clark refractor in 1869, he began his study of the heavens in earnest and published his first list of 81 new doubles in 1870.

His work and numerous discoveries opened doors for him throughout the world of astronomy. Burnham soon found himself welcome at observatories around the United States. During the course of his 42 years of observing, Burnham discovered over 1,200 doubles. It is a compliment to his keen-sightedness that many of the pairs he discovered and measured with his 6-inch Alvan Clark are now considered test objects for much larger instruments. In 1906 Burnham compiled his recently completed observations using the 36-inch refractor at Lick Observatory into the two-volume publication, *General Catalog of Double Stars*. This massive catalog, containing over 13,000 measurements, set the standard for double-star studies until the publication of Aitken's catalog in 1932, which contained over 17,000 pairs.

Robert Aitken entered double-star astronomy in 1899 with a systematic study of stars listed in the *Bonner Durchmusterung* (BD) catalog of stars. Compiled in 1855 at the Bonn Observatory, the BD listed 458,000 stars. Aitken's project was a truly massive undertaking. For the next six years Aitken and Professor W. J. Hussey worked on the project. In 1905 Hussey left the project and Aitken was forced to continue alone. By the time the project was completed and the final catalog published in 1932, Aitken had added 3,105 more pairs to the known list of double stars. In his six-year tenure with the project, Hussey discovered 1,329 doubles. This massive catalog is known as the *Aitken Double Star Catalog*, or simply the ADS.

The current, authoritative listing of double stars was published in 1984. The *Washington Double Star Catalog* (WDS) contains computerized information on 73,000 pairs!

As you can see, in order to be considered an astronomer during the nineteenth century, you had to study double stars. These studies helped broaden our scientific knowledge by leaps and bounds. From this early work came the subsequent discovery of the masses of objects outside our Solar System. It gave us an understanding of the dynamics of stars and stellar systems. It was demanding and difficult work contributed by great men of science, both amateur and professional.

THE LANGUAGE OF DOUBLE STARS

Of the 400 billion stars that make up the Milky Way galaxy we call home, 25% of them are double stars. Called *physical pairs*, they are stars bound together by gravity that orbit around a common center. Physical pairs may be a *visual binary* system in which the component stars are visible to us. Stars like Mizar, Alpha Centuri, and Sirius are visual binaries. They may have a relatively dim companion, like Sirius and Sirius-A, but they are capable of being resolved by optical telescopes.

Another class of binary is the *spectroscopic binary*, which is composed of two stars so close together that the only way to detect the companion is by an analysis of the light reaching Earth. A final type of binary, the *astrometric binary* system, is composed of a dark body circling the primary star. This type is detected by the motion of the star as it moves across the heavens. If another civilization were looking at our Sun, they could tell that there were planets circling it because of the irregular path the Sun would scribe against the stars over 50 years of observation. In another type of double star, an *optical pair*, the stars simply happen to lie along our line of sight from Earth and have no relation to each other.

Of these groups, the only ones that need concern amateurs like us are visual binaries and optical pairs. Studying these stars gives us a chance to follow in the footsteps of some of astronomy's great observers and to compare our visual acuity with theirs.

Regardless of the depth of your interest in double stars, there are a number of terms you need to know and understand before starting any observational work.

The work of the observers we have just examined is reflected in the stars. The objects recorded in the catalogs of these observers retain designations that note their place in the particular catalog and their discoverer. For example, Gamma Andromedae is entered as the 205th object of F.G.W. Struve's catalog. As such, it is preceded by the capital Greek letter sigma (Σ). The star is therefore also known as Σ 205.

Other designations found on star charts and in catalogs used by observers are:

OΣ	Otto Struve
O$\Sigma\Sigma$	Otto Struve, later catalog
H	William Herschel
h	John Herschel
Hh	J. Herschel's catalog of William's discoveries
A	Robert Aitken
ADS	*Aitken's New General Catalog of Double Stars*
β	S.W. Burnham
I	Robert T. Innes
Ho	G.W. Hough
Hu	W.J. Hussey

The stars that comprise the double are usually labeled A and B, with the brighter being A. In the case of a multiple star, the stars are labeled A, B, and C. During the golden age of the double star, any dim stars seen in the field of a double star were assumed to be components of the double star and, as such, were referred to as *comes* (Fig. 11-1). In many cases, however, these comes were optical illusions lying along our line of sight toward the double. An example of this would be the star Aldebaran. It is a true double star with a dim, magnitude 13 companion 31.4 arcseconds away. In the same field, however, is a magnitude 11 star at 121 arcseconds at PA 34 degrees. This star is unrelated to Aldebaran and only shares a common line of sight when seen from Earth.

The brighter star is also used as the primary in the measurement of the *position angle* (PA) of the pair. Expressed in degrees, the PA is measured counterclockwise starting from the north. If the component is west of the primary it is said to be *preceding*. If it is east of the primary it is said to be *following*.

The separation of the pair is measured in terms of arcseconds. Optical doubles usually have a separation of 30 arcseconds or more. In many cases, however, two stars can exceed 30 arcseconds and still be a common pair. Beta Cygni, separated by 34 arcseconds, is believed to be a physical pair, although no movement has been noticed since its discovery in 1832 by F.G.W. Struve. If two observed stars are separated by over 40 arcseconds it is almost a certainty that they are unrelated.

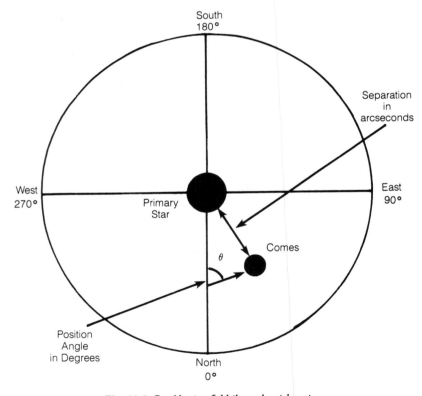

Fig. 11-1. Double star field through a telescope.

OBSERVING DOUBLE STARS

The observation of a double star is dependent on a number of factors. The ability of the observer's instrument to separate the two points of light that make up the double star is called the *resolving power* of the telescope. It is impossible for any telescope to make the disk of a distant star visible. It does not matter if you are using a small telescope in your backyard or the Hubble Space Telescope. There is no optical instrument that can show the real disk of a star because it is too small. What you see if you throw your eyepiece slightly out of focus is our old friend the airy disk. If two airy disks are close together, they can merge into a single image. The image is considered resolved only when you can recognize those two separate disks. Resolution limits for common amateur telescopes were listed in chapter 2.

The resolution limit of your telescope is set by the aperture only. Magnification has nothing to do with it. If you are observing with a 60mm, you will never be able to split a double that is 1.5 arcseconds apart no matter how good the seeing or how high the magnification. So if you are the proud owner of a 60mm refractor and you head outside tonight, you will be able to see double stars down to your instrument's dawes limit of 1.9 arcseconds. Right? Nope, not even close!

Remember that these limits are only theoretical, in practice they are rarely reached. One reason for this is that both the rayleigh and dawes limits are based on a pair of stars of equal or nearly equal brightness. When you are trying to resolve an unequal pair, with one star of magnitude 3 and the companion at magnitude 9, there is no real problem if it is a wide pair. If the stars are close, however, the dimmer companion can become lost in the glare of the brighter star.

The best example of an unequal pair is the companion of Sirius. Listed as a magnitude 7 star at 10 arcseconds distance from the primary, Sirius-B would be visible if it circled any star except the one it does. According to our chart of instruments and their limits, Sirius-B should be easily visible in a 60mm refractor. It isn't because the brightness of Sirius literally drowns out its fainter companion. Theorized by the irregularities in the motion of Sirius, Sirius-B was discovered in 1862 by Alvan Clark as he tested his new 18.5-inch refractor. This telescope was later purchased by the newly formed Chicago Astronomical Society, which later donated it to the Dearborn Observatory at Northwestern University, where it is still in operation.

Another problem comes when observing very faint pairs. It might be impossible for the eye to get enough contrast between the airy disks and the space between them to allow the observer to see the separation. Each instrument and each observer is different. Few of us are blessed with the sight of a Burnham or the optical excellence of an Alvan Clark telescope. The observer and his instrument form a unique personal equation that must be taken into consideration when trying to determine an observing session. The best solution would be for each individual to compile a scale that would determine exactly what pairs he could or could not split.

One way to measure this personal equation is for the observer to construct a peterson diagram. Named for Colorado amateur Harold Peterson who developed it in the early 1950s, the diagram is an excellent way to determine how your telescope will perform on double stars. Working under the transparent skies of Durango, Colorado, Peterson determined that the main factor in resolving faint or unequal pairs is not the magnitude difference between the two stars, but the difference between the fainter component and the limiting magnitude of the instrument being used.

With this theory, Peterson used a 3-inch (75mm) refractor with a 25mm eyepiece to study 125 double stars. As he examined each double, he noted whether it resolved or not. Later he recorded his observations on a chart, plotting magnitude along the X-axis and separation along the Y-axis. Resolved pairs were recorded as filled-in circles and unresolved pairs as open circles. His completed chart showed that his telescope could resolve pairs with a magnitude 9 companion as close as 3 arcseconds (Fig. 11-2). After that magnitude limit was reached, the detection of fainter companions required wider separation.

In order to get an idea of the resolution limit of your own instrument, it is a good idea to create your own peterson diagram. First, determine the limiting magnitude of your telescope. Don't go by the published numbers. Take the scope out and determine the faintest star you can see with it. Use an AAVSO chart like the one in chapter 7 as a guide. For uniformity, use the same eyepiece for all observations.

Observations should be made when seeing is good. Peterson used the visibility of the companion of Polaris as a gauge of seeing. Doubles should be well placed, preferably on or near the meridian. Using a list of doubles, record whether or not you can split the double. When you have accumulated a number of observations, about 100 should suffice, transfer your results to a graph. With your own diagram complete, you will know instantly if a double star is within reach of your telescope.

Once you have determined the peterson limit of your instrument, it is only human nature to try to exceed it. Of the three limits we have discussed here—the rayleigh, the

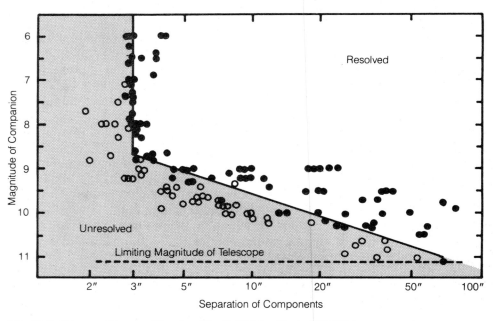

Fig. 11-2. Peterson diagram. (Courtesy Sky Publishing Corp., ©1980.)

dawes, and the peterson—the first two are based on the unchangeable fact of instrument aperture and cannot be exceeded. The peterson limit is a standard drawn against one set of circumstances. It assumes that the observations you made to construct the chart were made at your normal observation site. It also assumes that all observations were made with the same eyepiece, one that did not give a very high magnification. Because the peterson limit is a function of the telescope's limiting magnitude and the magnitude of the fainter member of the double-star system, you can try to exceed it and you may succeed.

There are a number of strategies you can use when trying to split a very close double. One of the easiest is the use of an occulting bar at the field stop of the eyepiece (Fig.

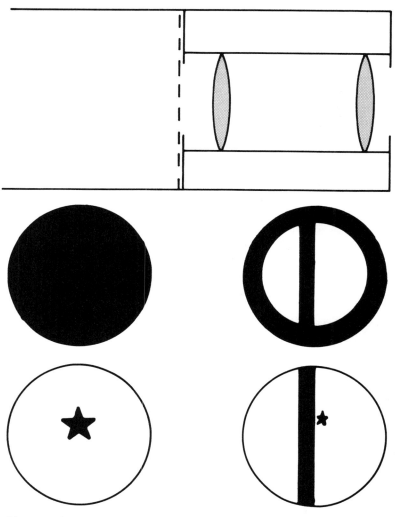

Fig. 11-3. Occulting bar.

11-3). To construct one you need a strip of thin, dark material that you can cut or shape easily. I used a sheet of exposed film and cut a circle in it the size of the inside of my eyepiece barrel. I then shaped it until I had a small strip 25mm long and about 5mm wide. Finally, I cemented this to the field stop of the eyepiece. To use the bar rotate the eyepiece until the brighter component is hidden behind the strip. The fainter, close companion should become visible when the glare of the primary star is cut by the film.

If you are using a refractor, another way to get a look at the close, dim companion is to make use of the diffraction pattern reflector owners take for granted. The pattern is created when light passes around the four vanes that hold a reflector's secondary mirror in place. It shows up as four *spikes* that are seen around the star. The light from the star is channeled into the spikes, leaving four spaces in which a dim object can be seen. Because the refractor has an unobstructed light path, you have to place something in the way of the light to get the diffraction pattern such as a diffraction mask. The easiest way is to criss-cross two pieces of duct or electrical tape across the end of the dewcap. *Do not place the tape on the lens*!

Now the light entering the tube will be diffracted by the tape, and you will see four spikes around the brighter component. The only drawback to this method is that the tape cannot be moved around easily, so the double components must be placed in such a way that the fainter star is in the space created by the diffraction pattern. You can remedy this by using some ingenuity and a piece of cardboard or plastic to make a sleeve that fits over the end of the dewcap to which you can attach the tape.

Some observers recommend a tape width of 10mm for acceptable results, but I think that it would be best for each observer to determine his or her own tape size. I cannot use a 10mm band, but find a size closer to 20mm sufficient. Make up a variety of strips and try them on various types of doubles. Is there a difference when using a wide band on a very close double? A narrow band? Keep track—nothing is ever a waste of time in science and astronomy is no exception.

A similar method makes use of a special type of aperture mask (Fig. 11-4). Using tape, film, or even cardboard, create a hexagonal (six-sided) mask to fit over the end of the dewcap or telescope tube. Use a sleeve so that you can rotate the mask. This hexagonal mask will give you a diffraction pattern of six spikes around a bright star. The sleeve allows the pattern to be rotated and so that the faint, close companion is easy to locate in the glare of the primary. Like the four-spike cross, it is best to experiment with a variety of sizes for the mask.

 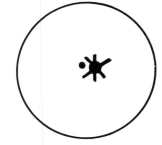

Fig. 11-4. Aperture mask.

The average eye can resolve a pair of stars 4 arcminutes apart. A star like Epsilon Lyrae, at slightly over 3 arcminutes in separation, appears as an elongated star to the naked eye of most observers, although the sharp-eyed can distinguish the two stars. Looking at Epsilon with a pair of 7×35 binoculars clears up any doubt as to its duplicity: the two stars stand clearly separate.

The components that make up the Epsilon Lyrae "double double" are much more difficult to separate. Epsilon-1 is the wider of the two systems with a separation of 2.8 arcseconds. Epsilon-2 is separated by 2.2 arcseconds. Both stars are well within the limits of both a 60mm and 80mm refractor. Also, they transit the meridian high in the sky, and the stars are all about equal in brightness. All of these things combine to create a unique test object for the small telescope. On a night with good seeing, begin with a low-power eyepiece and work your way up in magnification. Which eyepiece allows you to see the elongated images of both stars? Which eyepiece actually resolves Epsilon-1? Which resolves Epsilon-2? With an 80mm and good seeing, I can resolve Epsilon-1 with a 12mm ortho. Epsilon-2 takes a bit more. I find a 9mm sufficient for the entire four-star system.

When trying to resolve a close pair, you must remember that the eye can resolve two objects 240 arcseconds apart with no assistance. Using 240 arcseconds as a threshold of visibility, divide that figure by the separation distance of the pair to determine an approximate magnification for resolution. At 2.7 arcseconds, Epsilon-1 Lyrae should need a magnification of about 88 for resolution; Epsilon-2, about 110. An object like Gamma Arietis with a separation of 7.8 arcseconds needs a magnification of about 30. Use this formula as a guide, but remember a guide is all it is.

Magnification is dependent on both stable seeing and a solid mount. At high magnification, the field of the telescope seems to speed past your eye so fast that it is disorienting. Remember that, as the telescope magnifies an image, it also seems to increase the speed at which the Earth rotates. You must have a very solid mounting to be able to concentrate on a pair at high power. The best mounting is an equatorial equipped with a drive, but a smooth altazimuth can be used with a steady hand.

Observers in the eighteenth and nineteenth centuries were fascinated by double stars and the colors of the components. The colors described by Admiral Smyth, Rev. Webb, and other observers were so beautiful that they brought tears to the eyes of some of their readers. Imagine the contrasting colors of apple green and cherry red given by Smyth for 95 Herculis! The elder Struve went as far as developing his own color scale, including what he called pinkish-green, when recording his observations. These astronomers thought that by observing the colors of the stars they could gain insight into their makeup. For many years these reports were scoffed at and written off as overactive imaginations on the part of the observers. Now we are finding that they were partially correct.

In many cases, the colors we see when we observe a pair are caused by the reaction of the eye's rods and cones to the two stars. But there does seem to be a correlation between reported colors and spectral type. Research into this subject was conducted by four amateurs using the 24-inch reflector at the Mt. Cuba Observatory in Delaware. After studying some 200 pairs, they found a strong correlation between color, spectral type, and surface temperature. They further found that pairs with inconsistent color reports were those with a wide separation. The closer the pair, the more consistent the color reports.

During any observation it is therefore essential that you record the apparent colors of the stars you are observing. Not only are the colors pleasing to the eye, but they can tell you quite a bit about the makeup of the pair under observation. Stars that appear red, for example, are usually members of spectral class M, while stars seen as blue or with blue tints belong to spectral classes O through A. Remember, these conclusions were reached by amateur astronomers like you! They had access to professional equipment, but they gained that access by taking their astronomy seriously. So don't let anyone tell you that the days of serious contributions by amateurs are over!

Making Measurements of Double Stars

Okay, you are interested in observing double stars, but what can you do other than look at them? For a start, you can measure their separations and estimate their PA. "Who, me?" you respond. "I don't have one of those fancy micrometers and I don't have the money to buy one." Don't worry, my friend, most of us don't have access to precision micrometers either, but there are other ways to measure doubles. Interested?

The easiest way to make measurements is with a micrometer reticle eyepiece (Fig. 11-5). Available from a wide variety of companies, one can cost between $60 and $150. In addition to its use with double stars, it can help measure planetary disk size and determine CM for transit timings, so it's a handy item to have around. The reticle eyepiece is built around a piece of optical glass that has a cross-line pattern scribed onto it. The reticle scribed on the glass is broken into a number of major divisions, which are further divided into a number of minor subdivisions (Fig. 11-6). The glass fits onto an eyepiece, usually from 12mm to 6mm in focal length, at its focal plane. Most of these reticles include a red LED that illuminates the reticle scale and makes it easy to see against the dark background of the sky.

Fig. 11-5. Meade micrometer reticle eyepiece (MRE). (Courtesy Mark Stauffer.)

Before you use the reticle eyepiece, you have to calibrate its subdivisions. It is helpful to know just how many arcseconds each one covers. Mark Stauffer, an amateur living in Baton Rouge, Louisiana, uses a reticle eyepiece and recommends the use of a wide double for this calibration. Using a Meade reticle eyepiece with a focal length of 12mm on his Celestron C90 telescope, Mark has determined that each of the divisions scribed into the reticle is equal to 8 arcseconds. This figure is going to change for each telescope, so if you want to use a reticle eyepiece you must calibrate it for your own system. Mark uses Epsilon Lyrae, which has a wide separation of 208 arcseconds.

To do the calibrations, adjust the reticle so that the micrometer line passes through both major components, then simply count the number of divisions between them. For Mark's C90 this worked out to 26 divisions. If you divide the measured distance in arcseconds (208) by the number of divisions (26), you get 208/26 or 8 arcseconds per division. This will be different for each telescope so check yours.

Mark further tested his reticle by measuring the separation of the double star Gamma Arietis with the two stars covering one division of the reticle. This gave an observed separation of 8 arcseconds because the actual separation is 7.8 arcseconds. Not bad, eh? A micrometer reticle eyepiece is not able, however, to tell you the PA of the pair. This will have to be estimated by the eye.

Another, more accurate device for determining the measurement of a double star is a diffraction grating micrometer (DGM). The principle behind the DGM is the fact that when the image of a star passes through a series of closely aligned slits it is seen as a bright point source surrounded by a number of satellite images (Fig. 11-7). The total width of the bar and slit (p) can be changed so that you can vary the distance between the primary and satellite images (z). By varying z, which will be different for each observer, a variety of gratings can be made, each tailored for a specific separation distance (Fig. 11-8). The z value used for a pair should be close to, but not larger than, their actual separation. Because each person has a different z value, see Fig. 11-9 for a method you can use

Fig. 11-6. The MRE reticle.
(Courtesy Mark Stauffer.)

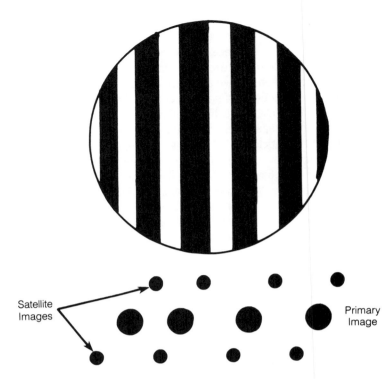

Fig. 11-7. The DGM grating and the type of image it produces.

Satellite Images

Primary Image

Fig. 11-8. Various size spacings for the diffraction DGM. (Courtesy Mark Stauffer.)

to determine yours. The actual measurements of the double star are made by rotating the grating so that the images of the stars form a right triangle. A reading of the angles involved is made from the micrometer (Fig. 11-10). The separation and PA can be computed from the average of a number of measurements. A DGM can be constructed at very little expense and can be used with any type of telescope. Mark Stauffer built and uses a DGM with his 80mm f/11.4 refractor and finds the results a great improvement over the micrometer reticle eyepiece (Fig. 11-11).

To make a DGM you will need some poster board, a couple of protractors, and a sharp knife. Cut a circle from black poster board that corresponds to the diameter of the telescope's dew cap. Then draw a series of lines that correspond to the p value you selected. Next fit two protractors onto the edge of the completed grating, aligning the zero point of the protractors to the horizontal position of the grating (Fig. 11-12). The appearance of the unit can be improved by making the protractor/micrometer scale fit flush along the barrel of the dew cap (Fig. 11-13). This can be done easily by using tape, a ruler, and graph paper. Mark mounted his DGM by making a ring from a 1-inch-wide-by-1/8-inch-thick aluminum strip. He used the lip from a margarine tub lid to attach the entire assembly to the telescope (Fig. 11-14).

The accuracy of the DGM depends on the separation of the pairs being measured. The ability of the telescope to resolve close pairs is crucial to the accuracy of the DGM because it can give a built-in error factor to the measurements attained with any size

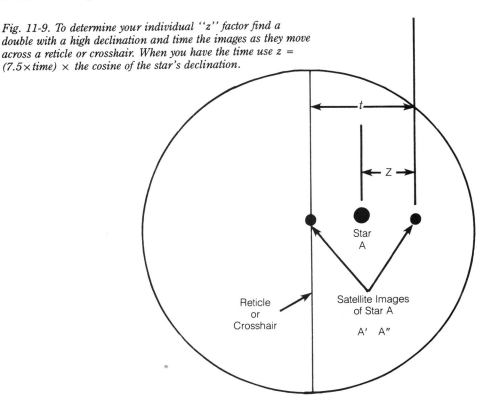

Fig. 11-9. To determine your individual "z" factor find a double with a high declination and time the images as they move across a reticle or crosshair. When you have the time use z = (7.5×time) × the cosine of the star's declination.

Fig. 11-10. The DGM can be adapted to any telescope. This DGM is attached to a 6-inch reflector. (Courtesy Mark Stauffer.)

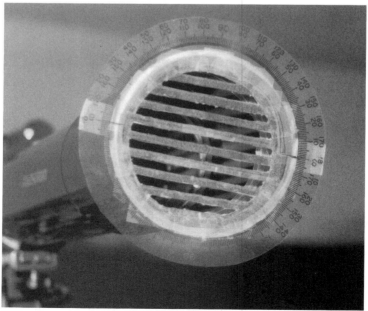

Fig. 11-11. Diffraction grating micrometer (DGM) on an 80mm refractor. (Courtesy Mark Stauffer.)

Fig. 11-12. The bar grating and protractors mounted to serve as the micrometer. (Courtesy Mark Stauffer.)

Fig. 11-13. The protractors can be replaced by a laminated scale for more accuracy. (Courtesy Mark Stauffer.)

Fig. 11-14. DGM on 80mm refractor. (Courtesy Mark Stauffer.)

telescope. When he used the DGM with his 80mm telescope on close pairs, Mark determined that the DGM is subject to theoretical errors of up to 30%. As the separation of the pairs increases there is a corresponding decrease in the theoretical error to about 7%. This means that you can more accurately measure wider pairs. In actual use Mark determined that for pairs of about 10 arcseconds separation he had an error of 3%, while for pairs of 5 arcseconds separation the error increased to 10%.

So what do you do with this DGM now that you've made it? Well, an excellent project would be to examine and measure some of the wider pairs listed in a catalog like the second volume of *Sky Catalogue 2000*. Professional astronomers are interested in the properties of doubles, but they tend to limit their interest to very close visual pairs, spectroscopic binaries, and astrometric binaries. Their interest in wide visual pairs is limited to taking measurements, and computing their orbits and possibly an ephemeris containing predicted separations and position angles. There is very little follow-up on the accuracy of these predictions, especially for wide pairs. An amateur using a DGM could measure and test this accuracy very easily.

DOUBLE STARS FOR THE TELESCOPE

There are so many double stars in the sky that it is difficult to come up with an all-inclusive list. Over the years Mark Stauffer has come up with a list of double stars that he feels shows the wide variety of colors and separations found in these objects. All of his observations were made with an 80mm f/11.4 refractor. Mark's list is a good starting point for your own observations because each of the stars is easy to find. Mark's comments follow each entry.

Beta Cygni	(Alberio) Gold and blue; best seen at low power. One of the finest pairs in the sky!
Gamma Andromedae	A fine yellow and blue double. Just as nice as, if not nicer than, Alberio! Best viewed at moderate magnification.
Gamma Arietis	A lovely pair of equal magnitude stars that look like a distant pair of headlights.
Epsilon Lyrae	The renowned "double double." I've observed this remarkable pair of pairs at 105× and was able to split both pairs of stars, provided the seeing was good.
Zeta Ursa Majoris	(Mizar) An easy pair of magnitude 2 and 4 stars that resolve at 35×. A very nice system and field too, with Alcor nearby at about 12 arcminutes. A faint star is located between Mizar and Alcor off to the side.
Alpha Ursa Minoris	(Polaris) A nice unequal pair at magnitudes 2 and 9 separated by 18 arcseconds. Look very carefully for the bluish-colored comes.
Xi Bootes	A wonderful color-contrasted double that is also a binary star. The primary is yellowish while the companion sometimes appears orange, sometimes reddish-purple. A fine pair to observe at moderate to high power.
Beta Orionis	(Rigel) A fine unequal pair at magnitude 0.1 and 6.8. The companion is easily observed at 105×.
Gamma Delphini	An easy pair of yellow stars about 10 arcseconds apart. Close by is another pair (Struve 2725), which are resolvable at 35× when the seeing is good.
Zeta Lyrae	A very nice contrasting pair of yellow and blue suns. An easy pair at low power.
Omicron-1 Cygni	A very wide color-contrasted triple with one companion at 107 arcseconds and the other at 338. Both are most likely optical companions. Can you tell that one of the companions is bluer than the other? Which one is it? A really nice sight at low power in a small scope.
Lambda Orionis	A close pair of yellow and light blue stars worth viewing at moderate to high magnifications.
Theta Orionis	(The Trapezium) An excellent quadruple found within the Orion Nebula (M42). An easy target for small telescopes at any magnification.
Epsilon Bootes	A difficult pair for the small scope, but I was able to split the yellow-orange primary and greenish companion with 182×.
Iota Cassiopeiae	A fine close triple of magnitude 5, 7, and 8 stars. The magnitude 5 and 7 stars are separated by 7 arcseconds, while the magnitude 5 and 8 stars are 2 arcseconds apart. Best observed at high magnifications. Some color to this system: white primary, bluish companions.
Eta Cassiopeiae	A real jewel of the night sky! A binary with a yellowish primary and a companion whose color I've seen to be orange, red,

	even purple! A double worth observing at any magnification.
Epsilon Pegasi	(Enif) A very wide pair of orange and blue stars. A fine object to view at low power.
Alpha Herculis	(Rasalgeth) A wonderful double composed of orange and greenish-blue stars separated by about 5 arcseconds. Best observed at moderate and higher power.
Beta Monocerotis	A very nice triple. The two companions themselves form a close duo. The entire triple can be best seen at moderately high power to take advantage of the close pair formed by the two companions.
Alpha Geminorum	(Castor) A truly fine binary composed of magnitude 2 and 3 stars, just resolved at $105\times$.
Gamma Leonis	A close pair of yellow stars that is within reach of a small telescope at moderate magnification. A pretty sight!

Observing
Variable Stars

*...approach the observing of variable stars with the utmost
caution. It is easy to become an addict...*

Leslie Peltier

That is probably the most accurate statement you will ever hear in amateur astronomy.
Variable star observing is habit forming. Variable star observers are among the most
active and dedicated members of the amateur astronomical community. In 1987, 521
observers in 36 countries contributed 264,566 observations of variable star magnitude
estimates to the American Association of Variable Star Observers (AAVSO). That averages out to 500 estimates per member! While that is an average, the number of estimates submitted by individual members range from a low of 1 to a high of 13,682. This is
not the number made over a lifetime of observing. It is for one year—1987! I would say
that these people are very dedicated! Wouldn't you?

The reason so many observations were made by such a small group of observers is
simple. These people know that their observations are going to be shared with professional astronomers who need this information. Even the people who were only able to
contribute one observation know that the single observation is significant because it, all
by itself, fills a need.

TYPES OF VARIABLE STARS

The stars that instill so much dedication in their observers can change their brightness
for one of two reasons. If they change because something happens inside the star itself,
it is called an *intrinsic variable*. Stars that expand and contract to some internal rhythm,

stars that "hiccup" during their life cycle, and stars that literally blow themselves apart are all examples of intrinsic variables.

If the changes take place because an external event acts on the light reaching Earth, it is called an *extrinsic variable*. Examples are two stars that revolve around a common point so that one blocks the light headed toward Earth and stars covered with massive *starspots*.

Variables are identified by a letter designation. The first variable discovered in a constellation is always labeled *R* followed by *S*, then *T*, and so on, beginning over with *RR* when the last letter is reached. Variables also carry a six-number designation known as its *harvard designation*, which is simply the coordinates of the star for the year 1900. The first four numbers denote its right ascension in hours and minutes and the last two its declination. If the star is south of the celestial equator and will have a negative declination, the last two numbers are underlined or preceded by a minus sign.

For example, the first variable discovered in the constellation Cancer carries the name *R Cancri* and the harvard designation *081112*. The first variable discovered in Aquarius is *R Aquarii* and has the harvard designation of *2338-15*, denoting it is south of the celestial equator.

Each of these primary classes is broken into a number of variable subclasses. Each subclass includes stars that share similar brightness variation profiles, called *light curves*. The curves are a plot of brightness against the time it takes for the brightness to complete a cycle, or *period*. Most of these subclasses are named for the first or most famous star in the group.

Cepheid variables are named after Delta Cephei. They show a small brightness range of 1 to 2 magnitudes, and periods of from 1 to 70 days. Their light curve (Fig. 12-1) shows smooth growth and decline in brightness. Because they are subject to internal changes that cause them to expand and contract, these stars are members of the intrinsic variable group.

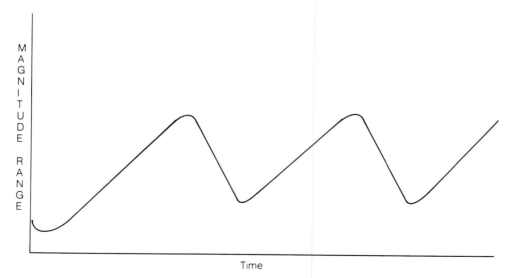

Fig. 12-1. *Light curve of Cepheid variable.*

RR Lyrae stars also pulsate, but they are much older stars than the cepheids. Named after the star RR Lyrae—catchy isn't it—their periods are much faster, never more than a bit over a day, and their brightness range is never more than two magnitudes. Their light curves (Fig. 12-2) show sharper peaks and steeper valleys than the smoother cepheids. Called *cluster variables*, many RR Lyrae stars are found in globular star clusters. They are members of the intrinsic group.

RV Tauri stars, named after the first star to be discovered with these characteristics, show as much as a three-magnitude range in brightness. Their periods, measured from one point of minimum brightness, the *minima*, to the next, can be from 50 to 150 days long. Their light curves (Fig. 12-3) show irregular brightness peaks and another minima, not quite as dim, between the deeper minima. These stars that behave so strangely are also members of the intrinsic group.

Long period variables (LPV) are the largest subclass within the intrinsic group. They have well-defined periods that last from 75 to 1,000 days and a brightness range that can exceed five magnitudes. Their light curves (Fig 12-4) show smooth rises to maximum brightness (maxima) followed by equally smooth drops to minima.

Related to the LPVs are the *semi-regular variables* (SR). These are stars that can't make up their minds. Some years they show regular maxima and minima, while the next year they show no variation in brightness. Another group of these indecisive stars are the *irregular variables*, which show little evidence of having any kind of period.

All of these variables have in common the fact that their brightness changes are caused by expansion and contraction, a pulsation of the stars themselves. These types of stars are called *pulsating variables*.

The next group of intrinsic variables we will meet have even more dramatic internal

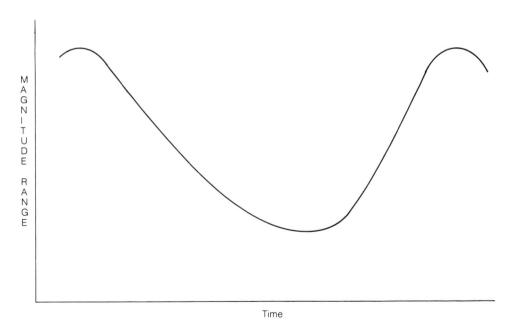

Fig. 12-2. Light curve of RR Lyrae variable.

Fig. 12-3. Light curve of RV Tauri variable. (Reprinted by special permission of the American Association of Variable Star Observers, through its Director Janet Mattei.)

Fig. 12-4(a). Long-period variable (LPV) stars can exhibit a variety of light curves. (Reprinted by special permission of the American Association of Variable Star Observers, through its Director Janet Mattei.)

Fig. 12-4(b).

0432+74   X CAM   MIRA

Fig. 12-4(c).

changes because they are all subject to massive eruptions of brightness caused when they blow themselves to pieces or just blow off their outer layers. Their handy designation is that of *eruptive variables*.

The extreme case of the eruptive variable is the *supernova*. A supernova can brighten by as much as 20 magnitudes, an increase of 90 million times its original brightness, in a very short period of time before gradually fading to a mere fraction of its original brightness, until it can't be detected at all.

The smaller version of the supernova is the much more common *nova*. Novae can brighten by seven or more magnitudes in a matter of days before they fade to their original magnitude over a long, even decline. Novae usually only brighten once during their lifetime, but another class, the *recurrent novae*, can undergo two or more outbursts during their life.

U Geminorum stars (U Gem) are also related to novae and recurrent novae. These stars can increase in brightness from two to six magnitudes, but they do so on a fairly regular basis at intervals of a few months. They return to their normal brightness much more rapidly, usually in about a week. Related to the U Gem stars are variables in the Z Camelopardalis family (Z Cam). Like the U Gem these stars are subject to rapid outbursts, but they can get stuck at a certain brightness as their brightness declines. These stars are also called *dwarf novae*.

UV Ceti stars, also called *flare stars*, also undergo extremely sudden and rapid increases in brightness of up to six magnitudes in a matter of minutes.

Finally we have the last of the intrinsics, the R Coronae Borealis stars (R CrB). Unlike the other eruptives, R CrB stars maintain a certain level and then decline in brightness as much as nine magnitudes before gradually returning to their former magnitude over a period of a few months to a number of years. They are sort of a reverse nova.

Eclipsing binaries are members of the extrinsic group. The cause of their variability is simple. The plane of the binary star system lies along our line of sight, so as star A and star B swing through their orbit, one blocks the light of the other and we see a drop in brightness at a regular interval. Beta Persei, also known as *Algol*, is an excellent example of such a system.

Although the amateur has a wide variety of variable types to choose from, it is easier for the new observer to get his or her feet wet by concentrating on long-period and semi-regular variables. So those are the types we are going to use here.

FINDING VARIABLE STARS

Finding variable stars is the hardest part of observing them. Most amateurs are used to looking for faint objects in their telescopes, but variables are different. When you are looking for a double star, your subconscious mind knows you are going to see two stars close together. The description of the double with its colors, separation, and position angle help the eye and the brain find the stars. When you are searching for a faint deep-sky object, your subconscious mind has the same advantage. It will give you certain visual clues about your target. It knows that whatever you are looking for has a certain size, shape, and configuration. It allows you to scan the star field and disregard the dozens of star images that don't fit the pattern.

So how do you visualize a variable star? It's not fuzzy. It's not a mass of stars, but a single star. It's not normally found with another star that is close by at a particular position angle. When you are looking for a variable star, there are no real visual clues. You are trying to pick one star out of a background of many. So how do you do this?

The first thing to do is learn about the general area through which you will be searching. AAVSO helps with this task by publishing a series of charts for each variable star (Fig. 12-5). On each chart is the basic information for the variable: its name and harvard designation, its magnitude range, period, and sometimes information about its spectral type or color. AAVSO charts come in seven versions, each keyed to a specific field of view size and each having a successively smaller area of coverage. The chart type is listed next to the box containing the star's harvard designation. Each chart contains the variable, which is usually indicated by an open circle, along with a series of stars and their magnitudes so that you can make brightness comparisons.

Type (a) charts cover an area of 15 square degrees on a scale of 1 millimeter (1mm) equal to 5 arcminutes. The limiting magnitude or faintest star shown is magnitude 7.5. These stars are listed and printed with north at the top of the chart. With this orientation, (a) charts are excellent for use with binoculars. You will notice that the charts list magnitudes as whole numbers. A magnitude 6.5 star is listed as 65 with the decimal point eliminated. This is so you won't confuse the decimal point with a star.

Type (ab) charts have a scale of 1mm equal to 8 arcseconds. They are issued for brighter stars and have a fainter limiting magnitude. Because they are also oriented with north at the top, they are useful for variables that can be seen with binoculars.

Type (b) charts cover an area of 3 square degrees and have a scale of 1mm equal to 1 arcminute. They contain stars as faint as magnitude 11 and are the first in the series to be printed with the correct astronomical orientation of south at the top. Type (b) charts are recommended for use with telescopes up to 3 inches in aperture.

Type (c) charts have a scale of 1mm equal to 40 arcseconds and are printed with south at the top. They are intended for use with telescopes from 3 inches to 6 inches in aperture.

Type (d) charts have a scale of 1mm equal to 20 arcseconds and are printed with south at the top. They are for telescopes 6 inches and larger.

Type (e) charts have a scale of 1mm equal to 10 arcseconds and are for use with faint variables. Designed for use with telescopes 6 inches and larger, they are printed with south at the top.

Type (f) charts have a scale of 1mm equal to 5 arcseconds. They are used when the variable is faint and the field contains many other faint stars.

AAVSO also publishes a number of *Finder Charts* for various areas of the sky. These cover a 10-square-degree area at a scale of 1mm equal to 6 arcminutes. They are designed to help locate the field of the variable and contain no magnitudes, so they are not to be used for making brightness estimates. A number of charts are also available for those with telescopes that have reversed fields, like refractors and catadioptrics with star diagonals.

You can use an (a) or (ab) chart to locate the target variable at the telescope and then use it or another chart to estimate the star's brightness. Not every variable on AAVSO's list is listed on a complete range of charts, however. Some of the brighter variables that

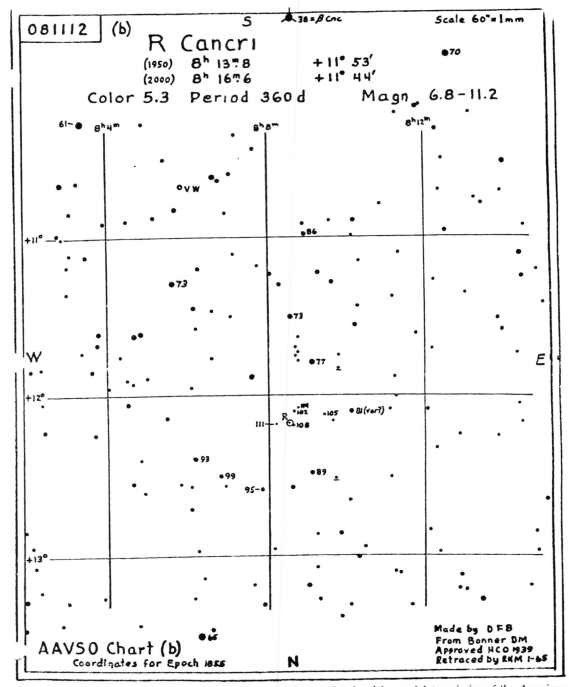

Fig. 12-5. *AAVSO variable star chart for R Cancri (R Cnc). (Reprinted by special permission of the American Association of Variable Star Observers, through its Director Janet Mattei.)*

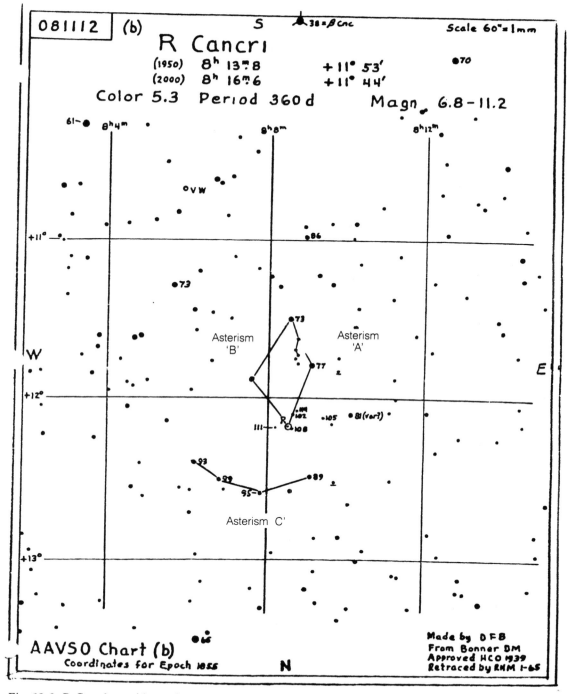

Fig. 12-6. R Cnc chart with asterisms marked. (Reprinted by special permission of the American Association of Variable Star Observers, through its Director Janet Mattei.)

can be seen through binoculars have only the (a) and (ab) charts, just as some of the fainter variables have only one or two charts.

Regardless of the number and type of charts available for your target variable, it is best to first locate the star in your own star atlas. If the star is too faint to be listed, use the coordinates given at the top of the chart and carefully plot the star yourself with a pencil. This lets you get a look at the big picture of the area in which the variable is located. Once you have the variable located or plotted, look around the variable for bright stars that will help you zero in on the variable. Use the ring you made with a coat hanger to get an idea of what the fields will look like in the finder.

Once you feel comfortable with the general area, take out the AAVSO charts for the variable and try to find visual clues that will help you find the star. In most cases there won't be any for you to use, so you are going to have to devise your own visual clues. Look on the chart for *asterisms* or groups of stars that form a recognizable shape. These will be your landmarks when you are at the eyepiece. For example, when I am trying to find R Cancri (Fig. 12-6), I look for three landmarks, especially when the star is faint. First I look for the little squiggly line of stars (A) just southeast of the variable. When I find this I know I'm close. Then I look for the variable to form a kite shape with three other stars (B) in the field. My final landmark is the arc of magnitude 9 stars (C) just north of the variable. This way I check my position not once but three times, so I'm sure I'm in the right place.

Once you have reached the area of the variable, you still must be careful. If you can see the variable in relation to the other stars, everything is fine. But when a variable is faint it can easily be confused with some of the comparison stars listed on your chart. Take some time to carefully check the position of each plotted comparison star. This might seem tedious, but it's better to be sure about which star is the variable than to make six magnitude estimates of the wrong star!

This entire process might put a damper on your enthusiasm for variable star observing before you start, but don't worry. The first time you add a variable to your observing list these precautions will make a lot of sense. You are venturing into a new area of the sky, an area that might be unfamiliar to you. The next time you observe the variable, your subconscious mind will know some landmarks because you've been over this ground before, and you'll find the variable a little faster. The next time you'll find it even faster, until you only have to point the telescope and the variable will be there, or so it will seem.

MAKING MAGNITUDE ESTIMATES

Now you are there. You've located the proper field, your eye is glued to the eyepiece, and you find yourself looking at a field of stars. What happens next? The desired end result of variable star observation is an accurate estimate of the star's brightness on the night of the observation or a *magnitude estimate*. To do this, simply compare the brightness of the variable with the brightness of known comparison stars that never vary their light output.

Remember all that tedious stuff about identifying each of the comparison stars? Well, this is where that pays off. The first thing you should do is identify the variable. Many variables, particularly the LPVs, appear red in color. That red color may be your

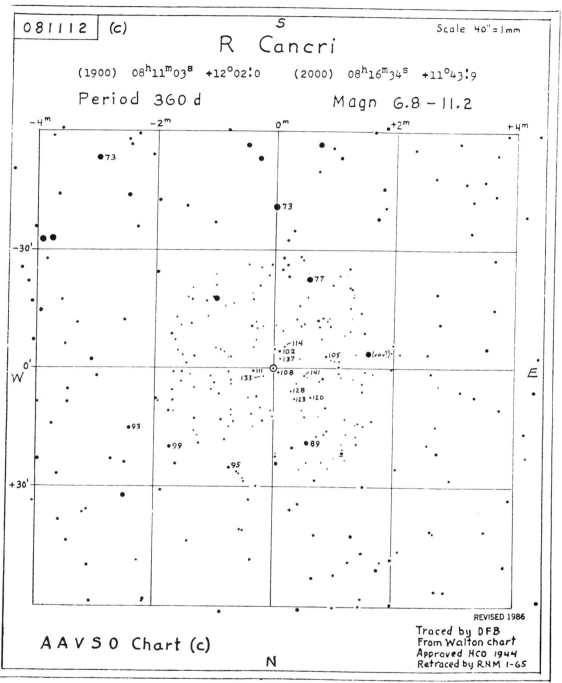

Fig. 12-7. Type C R Cnc chart. (Courtesy AAVSO.) (Reprinted by special permission of the American Association of Variable Star Observers, through its Director Janet Mattei.)

first clue to finding the right star, but don't expect the variable to pop out as a red star. Check the AAVSO chart and compare the positions of each star listed on the chart with what you see in the field of view. If the star is faint, you might have to switch to a higher-power eyepiece to bring out the fainter stars. If you do that, get out the next chart in the variable's series. If you use the (b) chart to zero in on R Cancri and can't quite make out which star matches the variable, switch to the (c) chart (Fig. 12-7). It will give you a better idea of how the other stars in the field are positioned around R Cancri and, if you've prepared well, you can use the asterisms you've marked on that chart to better refine the position of R Cancri.

If you have checked everything in the field and the variable still eludes you, chances are it is at a brightness below your telescope's limiting magnitude. Check the field once more and make a mental note, followed by a written one, as to which is the faintest star in the field that you can see. Indicate this with a notation like: *R Cnc not visible—(108.* This indicates that on the night of your observation R Cancri (R Cnc) was fainter than a star of magnitude 10.8, the faintest star you could see. The open parenthesis indicates fainter than and is a valid, and valuable, magnitude estimate.

If you find the variable, it is time to make a direct comparison. Let's say you find your target variable and it appears fairly bright. How do you proceed? First, decide which AAVSO chart you are going to use to make your comparisons and estimates. If the variable, in our case R Cancri, is very apparent in the field, you can safely use the (b) chart. Begin by letting your eye get used to the scene for a few minutes, then pick a star that you feel is close to R's brightness. Let's say it's that magnitude 93 star at the west end of the arc, star A in Fig. 12-8. Look at the two stars and quickly move your eye between them. Is R brighter or fainter than A? Fainter? Okay move north along the arc one star to star B at magnitude 99. Is R brighter or fainter than that one? Brighter? Good, you have determined that R Cancri is somewhere between magnitudes 93 and 99 (remember I'm leaving out the decimal point).

Are there any other stars in this magnitude range that you can compare with R? There sure is. Star C, the next star along the arc, is magnitude 95. Is R brighter or fainter than star C? After a few minutes you decide that R is slightly fainter than C, but brighter than B. Now it's time to compare all three stars—B, R, and C. Move your eye from B to R to C to R then back to B. Try to place where R's brightness falls. Does R appear to be closer to the brightness of one or the other? If it appears closer to C than to B, your estimate would be magnitude 96. If it appears to be closer to B than C, your estimate would be 98. If it appears to be midway between the two comparison stars, you would estimate R at magnitude 97. When you make your final estimate record it in your log. Our entry would look something like this: *R Cnc – 96. 93/95/99.*

This means that your estimate of R Cancri was magnitude 9.6 and that you used three stars for the comparison at magnitudes 9.3, 9.5, and 9.9. It is important to list the comparison stars you used, so be sure to mark them down.

The trick is to use as many comparison stars as you can to narrow the range of brightness. By doing this you are bringing the final estimate to within a narrow range of possibilities and improving its accuracy. If you find that there are only two comparison stars, use the two stars. But if you can, try to find a third, especially if you are making an estimate on a star for the first time.

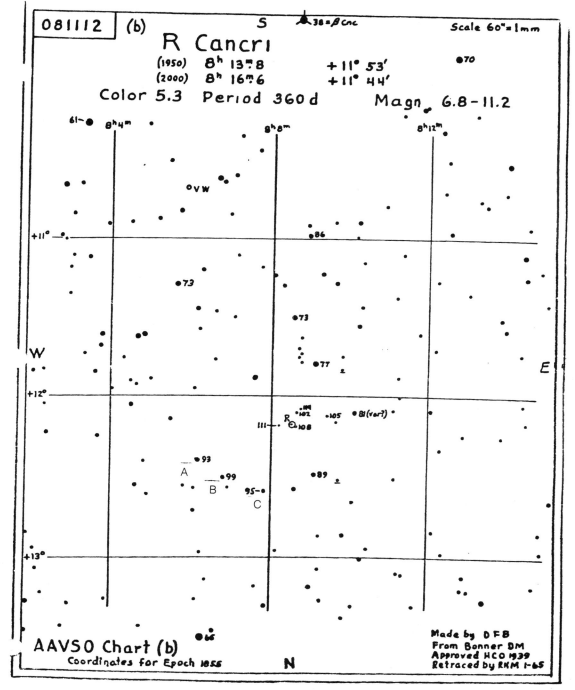

Fig. 12-8. Type B R Cnc chart with comparison stars marked. (Reprinted by special permission of the American Association of Variable Star Observers, through its Director Janet Mattei.)

You need to be careful when estimating the magnitudes of variables that appear to be red in color because of the *purkinje effect*. Red stars will excite the retina because the retina does not react to them as fast. (Remember dark adaptation and using a red light to protect it?) Because of this, the image builds up on the retina and causes a stronger signal to be sent to the brain than would other, nonred stars. Consequently red variables are seen as being slightly brighter. This can produce an incorrect magnitude estimate. When you are looking at a red, long-period variable try not to concentrate too hard on the star for any length of time. Keep your eyes moving around the field to minimize the amount of time the variable stays on any single spot on the retina. Another way to overcome the purkinje effect is to throw the field slightly out of focus to reduce the intensity of the red color.

KEEPING RECORDS

Once you have made all your estimates for the evening, it is time to put them in order. You want to use a system that keeps everything handy, but one that won't influence your estimates at the telescope. A good way to do this is to keep all your observations intact in your observation log, with one page for each night's work. Once inside transfer this information to a master log that allows one sheet of each variable star.

Looseleaf ledger sheets with six columns work well for this master log. You can head each sheet with information relevant to the variable and label the columns of the sheet with headings such as: remarks "the wide column", date, time (local), time (UT), estimate, comparison stars, and conditions. If you happen to have a personal computer, a spreadsheet program does this job nicely.

If you are submitting estimates to AAVSO, get in the habit of listing the date according to the Julian Day calendar. Julian days (JD) are listed consecutively from the starting date of January 1, 4713 B.C. July 31, 1988, for example, is JD 2,447,374. This gives AAVSO a scale against which to plot light curves and makes its predictions for variables.

Table 12-1. Conversion to Decimal Day for Observations Made Between.

EST	CST	MST	PST	Local Time — and — EST	CST	MST	PST	Decimal
324	224	---	---	548	448	---	---	0.4
549	449	349	249	811	711	611	511	0.5
812	712	612	512	1036	936	836	736	0.6
1037	937	837	737	1259	1159	1059	959	0.7
1300	1200	1100	1000	1524	1424	1324	1224	0.8
1525	1425	1325	1225	1747	1647	1547	1447	0.9
1748	1648	1548	1448	2012	1912	1812	1712	1.0
2013	1913	1813	1713	2235	2135	2035	1935	1.1
2236	2136	2036	1936	---	---	---	2200	1.2
---	---	---	2201	---	---	---	---	1.3

The above table is used to convert observation times of variable star estimates from local time to decimal fractions of a Julian Day. First convert the time of your observation to a 24-hour clock starting at noon. Three A.M. CST now becomes 1500 hours local time. Look at the table and see what two limits 1500 falls between (CST 1425 and 1647) for which the decimal day of your observation becomes 0.9. Add this figure to the Julian day of your observation. (Adapted from AAVSO, Janet Mattei, Director.)

The big difference is that JDs are measured from noon to noon instead of midnight to midnight, so if you insist on using our calendar, label observations made on the night of July 30 as July 30-31. AAVSO can supply you with the current year's JD calendar. A number of commercially available astronomical calendars list the JD as well as the date you are used to seeing. AAVSO also requires that observations be listed with the hour in decimal form. Table 12-1 gives you conversions for changing UT and local time into decimal hours.

To submit your estimates to AAVSO you need only fill out the organization's monthly form (Fig. 12-9). Reports should be sent in on the first of the month following the month being reported. For example, May estimates should be sent in at the beginning of June.

The form is rather straightforward, with the consecutive report number and the number of sheets in the report appearing on the first line. The month for which the report is being made goes on the second line and information on the observer and location is written on the next four lines. If you do not use UT indicate what you do use on line five and information on your instrument on line six. The variables should be listed in the ascending order of their harvard designation, followed by their letter designation, the JD, the decimal time, the magnitude estimate, the comparison stars used, and any remarks that you feel are necessary, such as ''10 degrees above horizon'' or ''high clouds''—anything that you think would have an influence on the listed estimate.

TELESCOPES FOR VARIABLE STAR OBSERVING

Variable star observing is the forum where amateur astronomers can really make their presence felt. You don't need a lot of expensive equipment. In fact, serious work can be done with a good pair of binoculars. If you are going to use binoculars, try to select variables that fade no fainter than magnitude 9. Appendix F lists a number of variables that have minima brighter than magnitude 10 and are suitable for observation with binoculars. If you don't have a telescope but are planning to purchase one, binoculars are a good way to learn the techniques and the discipline you'll need once you get a telescope.

Your observing list of variables will depend on the aperture of your telescope and its light-gathering power. It is a good idea for you to stay with variables that have minima just below the limiting magnitude of your telescope's aperture. Table 12-2 lists various instruments and the faintest variable you should try to observe with each one. A number of things can determine your variable threshold. The seeing conditions of your observing site and your own visual sensitivity might increase or decrease the lowest magnitude you can easily see.

One thing that also enters into the picture is the fact that not being able to see the variable is, in itself, a valuable observation. My 80mm (3.1-inch) refractor has a limiting magnitude of 11.5. This should limit the variables on my list to those with minima of magnitude 11.2 or brighter. I have included a number of variables that have minima below this limit, however, because of this fact.

When making your observations you will want to get the most out of your telescope's abilities. As with planetary observing, each type of instrument has its particular advantage. A short focal ratio system, like f/4 or f/5, will give you wide, bright fields of view that make it easy to identify variable star fields. Telescopes with longer focal ratios, f/8 and up, will have a smaller but darker field of view, giving you better contrast

THE AMERICAN ASSOCIATION OF VARIABLE STAR OBSERVERS
VARIABLE STAR OBSERVATIONS

Sheet __1__ of __1__ .   Report No. __3__

For Month of __MAY__ ____ Year __1988__

Observer __MICHAEL R. PORCELLINO__

Street _____

City __Chicago__ State __IL__

Country __USA__ ____ Zip Code _____

Time Used, (GMAT) or _____

Instrument (s) __80mm f/11.4 REFRACTOR__

AAVSO Observer Initials
__PLL__

For AAVSO HQ Use Only

Received _____

Entered _____

Verified _____

DESIGNATION	VARIABLE	JUL. DAY + DEC	MAGN.	COMP STARS	REMARKS
081112	R Cnc	2447287.6	(9.5		
081617	V Cnc	7287.6	78	72/78/82	
090151	V UMa	7287.7	(10.1		
072708	S CMi	.6	9.3	8.9/93/97	
094211	R Leo	.7	6.2	5.8/64/66	
122001	SS Vir	.7	8.5	83/86	
123307	R Vir	.7	(9.6		
1327-06	S Vir	.7	9.3	91/9.6	

TOTAL NUMBER INNER SANCTUMS*   TOTAL NUMBER OBSERVATIONS REPORTED __8__

*An Inner Sanctum is an observation which is magnitude 13.8 or fainter, or < 14.0 or fainter.

Observations should be sent to AAVSO Headquarters, 25 Birch Street, Cambridge, MA 02138, USA, as soon as possible after the first of each month.

Fig. 12-9. AAVSO report form.

Table 12-2. Various Size Amateur Instruments and Variable Star Limits

Aperture (in)		Limit of Variable	
EYE	10.0	6.2	14.5
1.4	12.5	9.5	15.4
2.0	13.1	10.3	15.5
2.4	14.0	10.7	15.6
3.1	16.0	11.3	16.0
4.0	17.5	11.8	16.1
6.0	20.0	13.0	16.4
8.0		13.6	

between the stars and the dark background. The important thing in variable star observing is not the focal ratio or even the aperture of the telescope, but the person using the telescope.

Another nice thing about variable star observing is that you don't need a bevy of eyepieces. With my 80mm I use a 26mm plossl, and an 18mm and 12mm orthoscopic. The 26mm gives me a field just over 1° across and is perfect for finding the star field I want. I use the 18mm to make most of my estimates and keep the 12mm in reserve if the star is faint. I find I rarely need the 12mm; the 18mm and even the 26mm are perfect for my situation.

Variable star observing will teach you the limitations of your instrument, be it large or small. It will show you how to stay within those limitations while still contributing to science. Telescopes and eyepieces aside, you do need a few things. The first is the discipline to go out and make your estimates at every opportunity. If you join AAVSO, and I recommend that you do, you will be issued charts on two or three variables to get you started. You will also receive a list of predictions for the stars in AAVSO's observing program.

This handy little publication tells you the approximate dates predicted by the AAVSO computer for the star's maxima and minima. It also tells you, when possible, whether the variable is brighter than magnitude 11 or, if the minima falls this low, fainter than 13.5. With this list you can plan your sessions at the telescope and expand your own list of variables as rapidly or slowly as you like. AAVSO recommends that you observe each variable on your list twice a month.

Finally, variable star observing is a pursuit that fits in easily with other observational specialties. Dan Troiani, assistant ALPO Mars recorder, has a primary interest in observing the Red Planet, but when it's in conjunction with the Sun, Dan turns his 10-inch on variables. Since he joined AAVSO in 1979 he's made over 16,000 magnitude estimates!

✳13✳

Observing
Deep-Sky Objects

*...we are...approaching more amazing regions and fresh scenes
will open upon us of inexpressible and awful grandeur.*

Rev. T. W. Webb

If the Solar System offers the observer an abundance of objects to study, the deep sky
offers a veritable plethora of objects for the amateur's telescope. The number of star
clusters, nebulae, and galaxies that lie within reach of even the small telescope can keep
the amateur at the eyepiece for a very long time.

Deep-sky objects (DSOs) fall into three basic categories. *Star clusters* are groups of
stars bound to each other by gravity that move through space together. A *nebula* is a
filmy sheet of glowing gas surrounding a star or star cluster. *Galaxies* are ''continents''
of stars, millions and even billions of light years away from us, that float through the universe.

Countless amateurs have devoted their observational time to studying these objects
because of the staggering beauty of their structure. They have even been able to make a
few contributions to science along the way.

TELESCOPES, IMAGE BRIGHTNESS, AND THE DEEP SKY

The observer of Solar System objects is looking at the bright image of a planet spread
over an extended area. The double star or variable star observer is looking at point
images of individual stars. But the type of object of interest to the deep-sky observer is
quite different. DSOs are usually faint conglomerations of stars, gas, and dust spread
over a relatively large area of the sky. They are classified as extended objects like planets, but their brightness is hundreds and even thousands of times fainter.

With any object you see, whether with the naked eye, binoculars, or a telescope, there is a limited amount of light reaching your eye. Because deep-sky objects cover a large area, the light available to your eye is spread evenly over an equally large area. As a result, you see a faint object.

When you view a DSO through a telescope, you are forcing the available light to cover a larger area. Because the image is magnified its brightness decreases. If you continue to increase the magnification of the image, you will make things worse. There is a magnification point at which there is too little available light and the image blends into the background and disappears.

The key to successfully observing deep-sky objects is to get the brightest image possible from these faint objects. The observer measures the brightness of each deep-sky object three ways: by its integrated magnitude, its normal brightness, and its telescopic brightness. The *integrated magnitude* assigned to DSOs is not quite the same as the magnitudes given for point-source objects like stars. Because a DSO is an extended object, the integrated magnitude assigned to it is the magnitude the object would have if all of its light was compacted, or integrated, into a single point source. The observer will find that as the size of the area that the object covers increases, it will become harder to see the object, even when a relatively high magnitude figure is listed for it.

The normal brightness of a DSO is the brightness of the object as viewed with the naked eye. Wait a minute! A magnitude 9 galaxy is not visible to the naked eye, so how is it possible to measure this normal brightness? The objective of a telescope or binoculars is an extension of the human eye and increases the amount of light available to the eye based on the objective's light-gathering power. As you increase the size of the objective, more light is passed to the eye allowing dimmer objects to be seen. When you use an eyepiece with an exit pupil (remember them?) that equals the size of the eye's pupil, your eye is able to use all the light that the objective is able to collect. This normal exit pupil allows the eye to sense this image across the entire retina so the telescopic brightness seen is the same as the object's normal brightness.

All this relates to the type of telescope best suited for observing these "faint fuz-

Aperture (mm)	(in)	Smallest Angular Diameter
35	1.4	1.0°
50	2.0	43.0'
60	2.4	33.0'
80	3.1	27.0'
100	4.0	21.0'
150	6.0	14.0'
200	8.0	11.0'
250	10.0	8.4'
317.5	12.5	6.8'
330	13.1	6.5'
355	14.0	6.1'
405	16.0	5.3'
444	17.5	5.0'
500	20.0	4.3'

Table 13-1. Contrast Limits for Instruments of Various Sizes

Fig. 13-1. 20-inch f/5 dobsonian telescope.

zies,'' as they are affectionately known. To achieve an exit pupil of 7mm with an f/11 telescope, you would need an eyepiece of almost 80mm focal length. That is an awful lot of glass! But with a telescope of f/4 you only need a 28mm focal length eyepiece to get a 7mm exit pupil. While 80mm eyepieces are available, they are hard to come by and very expensive. A 28mm eyepiece, on the other hand, is easily obtained and won't put your budget on the critical list. So the first requirement for a deep-sky instrument is that it have a short focal ratio, usually f/4 or f/5. The extremely short focal ratio instruments, f/4 and shorter, are called *rich field telescopes* (RFT).

The normal exit pupil also sets a limit on the object's maximum angular size so that there is still enough contrast for it to stand out from the background field of view (see Table 13-1). Objects that are magnified to a larger size than this limit begin to lose contrast and grow dimmer, until they fade into the background and disappear. Objects smaller than this limit will appear as bright patches in the field of view. The larger the aperture of the telescope, the more light that is gathered and sent to the eye, and the smaller this size limit becomes. This way a smaller DSO can be observed at maximum brightness. So the next requirement for a deep-sky telescope is to have a large aperture. Combining a large aperture and a short focal ratio in a telescope allows you to use a higher magnification on a deep-sky object and still maintain adequate contrast between the image and its background.

The aperture of the deep-sky observer's telescope is a matter of individual choice. Some observers are quite happy with a 4- or 6-inch RFT, while others crave the dimmer limiting magnitude and greater image contrast that a 13- or even 17-inch aperture gives them. With the introduction of thin-mirror technology and compact dobsonian mounts, many amateurs are opting for larger-sized instruments. Apertures in the 20-inch range are considered portable (Fig. 13-1) and easily handled by one person.

AIDS FOR THE DEEP-SKY OBSERVER

Deep-sky observers use a visual trick to help them see the faint fuzzies they chase through the night sky. *Averted vision* is an excellent way to improve your ability to see deep-sky objects. When you look directly at an object, most of the light falls on the center of the retina. This area, the *fovea centralis*, is composed primarily of cones that react only to bright light. When the light from a DSO falls on the fovea you will see only a dim image.

In order to use the more numerous rods that are sensitive to faint light, you must direct the image away from the fovea and onto the rods. To do so, simply look at the side of the field of view. When the image falls on the rods, it will become brighter and you will be able to see it much better. Using averted vision is much the same as using peripheral vision to detect something out of the corner of your eye. When using averted vision you look at one part of the field of view while directing your attention to another part. Getting used to using averted vision takes a little practice, but after awhile it becomes almost an automatic response to detecting faint objects.

If you are going to get serious about deep-sky observing, you are going to need a good star atlas. While Norton's is great for the beginning observer, something with more detail is needed in order to hunt down really faint DSOs. It is easier for the observer to find a faint galaxy in a field crowded with similar objects if he can pick out and identify an individual star, from which he can get his bearings. In order to do this the observer must have an atlas with a deep limiting magnitude.

A good rule of thumb is to use a chart and a finder that have about the same limiting magnitude. This means an atlas that has at least a limiting magnitude of 8. Norton's only goes down to magnitude 6, but Tirion's excellent *Sky Atlas 2000* goes to magnitude 8. The recently published *Uranometria 2000* lists stars down to magnitude 9.5 and easily fills the bill as *the* atlas for the deep-sky observer. An atlas designed specifically for the deep-sky observer is the *Sarna Deep-Sky Atlas*. Deep-sky objects of various types are listed on 102 star charts that have a scale of 1mm equal to 1 arcsecond. Compared to the scales of Norton's atlas, *Sky Atlas 2000*, and *Uranometria*, this is the largest-scale atlas available. *Sarna* is unique among atlases because it not only lists the right ascension and declination of its 254 objects, it also gives the observer simple directions for star hopping to the object using bright stars and easy-to-find asterisms.

Another crucial item for deep-sky observers is the finder. Because you are looking for objects that are very faint, you need to be able to pick your way across the sky to your target. Star hopping between bright stars and asterisms is an excellent way to do this, but you will need an instrument that has a limiting magnitude low enough to allow you to find your key asterisms on your way to that faint fuzzy. The majority of finders available today are either 6×30mm (objective 30mm, magnification 6×) or 8×50mm (objective 50mm, magnification 8×). A few companies even market finders with an 80mm objective! For most amateurs, unless you are going the 20-inch dobsonian route, a 6×30 or 8×50 finder is more than adequate.

Many finders sold today come with a right-angle star diagonal that eliminates the need to bend over in order to look through it. While this may be convenient, it can make things look pretty weird in the finder's field. Remember the image orientation of the standard astronomical telescope? A finder is subject to the same orientation, after all it is an astronomical telescope even though it doesn't boast a huge aperture. Putting a star diagonal on the finder is only going to confuse you by giving you an inverted and reversed image that will not match any of your star charts.

When you buy a finder telescope that comes with a right-angle finder, check its image orientation. If the field is reversed and inverted most manufacturers make an adapter that eliminates the right-angle feature and allows you to sight straight through the finder. If you want the right-angle feature, switch the inverted/reversed diagonal with an amici diagonal so that you will have the proper orientation in your finder field.

Deep-sky observing requires a few things from the sky that other areas of observation do not need. If you are observing bright objects like the Moon, planets, bright double stars, and even the brighter variables, you are going to be relatively unaffected by light pollution and can stay happily in your backyard or nearby field. The observer of the deep sky, however, needs all that extra light like he needs a wire brush to clean his mirror! The farther away one can get from the lights of a city, the better it is for the deep-sky observer.

This aversion to light also means that the deep-sky observer wants to avoid that old nemesis, the Moon. Deep-sky folk have little use for our natural satellite and plan their observing schedules to coincide with the new Moon whenever possible. Deep-sky observation also calls for the atmosphere to be exceptionally clear. When the stars stand out like sparkling diamonds against the velvet background of the sky and you can see easily stars of magnitude 6, it is a perfect night for observing the deep sky. Because objects like nebulae and galaxies are faint and extended, they show up best during conditions of high transparency.

For the deep-sky observer, deep sky means dark sky! The question that comes to mind is where can the deep-sky observer go? There are no deep-sky parks listed on any map I have seen. The thing for the deep-sky observer to do is bundle the telescope into the car and hit the road. Depending on the size of the city you are fleeing, a drive of between 40 and 100 miles should bring you to nice dark skies. If you are lucky, you might only have to drive for an hour or two to reach a dark site. You can make the round trip, observe, and be back home for breakfast.

Many amateurs make dark-sky weekends a family affair, bringing the entire clan along for a weekend of camping and observing in a nearby state park. If your family is not the outdoor type, you may be able to find a nearby motel that can act as a base camp. Some amateurs have found remote, quiet nooks off the busy interstates and return there time after time.

If the site you find is on private property, get permission before you set up your gear. There is nothing like hearing the low growl of a farmer's dog while you are reaching for that 56mm plossl to get more than your visual purple off and running. If you take the time to explain to the property owners what you are doing and offer them a look through your telescope, you will usually find them only too happy to have you around.

For the amateur stuck under the light-polluted skies of the city, there are ways to overcome this handicap. A number of companies are producing a line of filters especially designed to cut through light pollution and bring the deep sky to the city observer. Light pollution reduction (LPR) *filters* screen out the light from some types of sodium vapor lamps and minimize skyglow. Lumicon's deep-sky filter, for example, can improve city observing by filtering out light pollution and increasing the contrast of the deep-sky object.

Other deep-sky filters select particular wavelengths of light for transmission to the eye. Ultrahigh contrast (UHC) filters increase the overall contrast between faint planetary nebulae and the background sky. Oxygen III (O-III) filters select wavelengths of light that complement the UHC filter by allowing through wavelengths unique to planetary nebulae. Hydrogen-beta line filters (H-BETA) are sensitive to light common in reflection and emission type nebulae.

Because deep-sky telescopes are usually more portable than others, it's important to have a source of power for the drive unit, dew zappers, and other little electrical goodies that you may use. I don't recommend using the adapter that lets you tap into your car battery. I tried it one night, ran down the battery, and was stuck until the next morning. Marine batteries offer an excellent, portable source of power for your drive and electrical accessories.

No matter where or what you observe, you are going to be faced with one adversary that is more persistent than even light pollution. It will ruin your nights by spreading itself over your optics, your charts, and your glasses. It comes on perfectly clear, transparent evenings and nights when the seeing for planetary work is better than excellent. I'm talking about every amateur's friend—dew.

Dew might look pretty on the filmy strands of a spider's web or the petals of a flower, but it doesn't look good on glass! Dew on your optics can cut off the light entering your telescope just as surely as putting a bag over the lens. The light hitting the dew-covered objective scatters every which way, turning the telescope into a very expensive kaleidoscope.

Over the years, amateurs have come up with a wide variety of ways to defeat this enemy. Some use fans to circulate the air over the objective. Some use hair dryers to dry off the offending dew. Some, admitting defeat, simply pack up and head for home. The best way to treat dew on your optics is to be prepared for it.

The easiest way to combat dew is to have an adequate dew cap at the end of your tube. For refractors or catadioptrics, the length of this dew cap should be equivalent to the diameter of the primary. So for a 4-inch refractor the dew cap should be 4 inches

Fig. 13-2. Heat ropes protect the telescope's finder from dew. Switch box allows the observer to also heat the secondary mirror and eyepiece with heat ropes.

long. With reflectors, the tube length should extend at least 4 inches past the secondary support, preferably more. The trend today is to place *heat ropes*, insulated strips of nichrome wire, around the optical surfaces of the telescope. Running a very small current through the ropes heats them up and they transfer their heat to the optical surface. This slight increase in heat is all that is needed to prevent dew from forming (Fig. 13-2).

On a refractor or catadioptric, running a heat rope around the inside of the dew cap is recommended. With reflectors, a heat rope on the secondary mirror as well as along the inside of the tube near the primary mirror is recommended. You also can protect your finder from dew using the same technique.

Some amateurs have even devised heaters for their eyepieces, although the eyepiece is the only element of the optical system that can be removed and warmed by itself. If dew forms on the eyepiece, you can hold it under your car's heater to dry it off or use your own body heat to get the same result. Whatever you do to combat dew, if it forms and you can't heat it away *do not* wipe it off the optics. This could scratch the optics. While dew will eventually dissipate, scratches never go away.

Dew also will soak your charts and notebooks. Star atlases that come in individual sheets, like *Sky Atlas 2000* and *Sarna*, can be laminated in plastic to keep dew off. This won't work however, with book-type atlases like Norton's and *Uranometria*. The easiest way to protect these atlases is to place a sheet of stiff, transparent plastic over them so you can see the chart directly through the covering.

LETTERS, NUMBERS, AND LISTS

When you are just getting into deep-sky observing, it helps to have a list of easy-to-see objects. Probably the best list of deep-sky objects ever compiled was put together by a French comet hunter who had very little use for these objects because they were so easily confused with comets. Charles Messier began compiling his list in 1758 when he came across the Crab nebula while looking for a comet he had found two weeks earlier. The object he saw puzzled him because it "was a little like...the comet I had observed before, however, it was too bright."

In order to save himself and other comet hunters the trouble of misidentifying this "whitish light," Messier recorded its location on one of his charts. The list that Messier began that night was originally published in 1774 and included 45 deep-sky objects. In 1780 Messier published a supplement to his original list that included 23 more objects. In 1781 he added 35 new objects, 24 of which were discovered by a colleague named Pierre Mechain, and published the final version of his catalog containing 103 objects. In the 1920s, the French astronomer Camille Flammarion and American astronomer Dr. Helen S. Hogg added another seven deep-sky objects, the descriptions of which they had found in Messier's and Mechain's notes.

The *Messier Catalogue* that we know today includes a total of 110 objects. All of these objects are easily visible through amateur instruments. Known by the preface M, followed by their catalog number, these objects comprise the créme de la creme of deep-sky objects.

An amateur interested in deep-sky observing would be well served to undertake a program to find and record all 110 Messier objects. Such a project should be pursued at a comfortable pace. There is no time limit unless you choose to impose one. There is no sense in rushing through such a pleasing task. Each object should be observed with an eye toward the detail you can make out with your instrument. Keep your observations in your observing log and include a written description and sketch of each object. "Doing the Messiers" will further develop your observational skills. If you use the starhopping technique to locate these objects, you will come away from the project with an even better knowledge of the night sky and the constellations.

Once your appetite for the deep sky is whetted by the *Messier Catalogue*, where can you turn? During the late nineteenth century, J. L. E. Dreyer cataloged deep-sky objects down to magnitude 12 for almost the entire sky. The *New General Catalogue of Nebulae and Clusters of Stars* (NGC) lists over 8,000 deep-sky objects for the amateur to chase. In the 1970s astronomers from the University of Arizona updated the information available on these objects and published the *Revised New General Catalogue of Nonstellar Astronomical Objects* (RNGC). The RNGC includes the original object descriptions from Dreyer's NGC, information from the National Geographic Society and Palomar Observatory Sky Survey conducted in the 1950s and 1960s, and updated position information.

After completing a survey of all these objects, the next logical target for the deep-sky observer is to observe as many of the NGC objects as his telescope can reach. Twenty years ago it would have been unthinkable for an amateur to even consider a study of the complete NGC list. But today's large-aperture telescopes put such a study well within the reach of the dedicated deep-sky observer.

STAR CLUSTERS

Some of the most spectacular deep-sky targets for the amateur's telescope are star clusters. Some, like the Pleiades (M45) and the Great Hercules cluster (M13), are visible

Fig. 13-3. The Beehive cluster (M44) in Cancer. (Courtesy Walter Piorkowski, ©1988.)

through binoculars, while others require a telescope.

Star clusters fall into two basic categories. *Open clusters* are groups of stars bound by gravity that move through space together. They can range from loose conglomerations of stars like the Beehive cluster (M44) in Cancer (Fig. 13-3), to tightly packed groups containing hundreds of stars like the Double cluster in Perseus (Fig. 13-4). A problem that you will encounter when trying to observe some of these clusters is difficulty identifying them. Although bright clusters may be immediately apparent, fainter

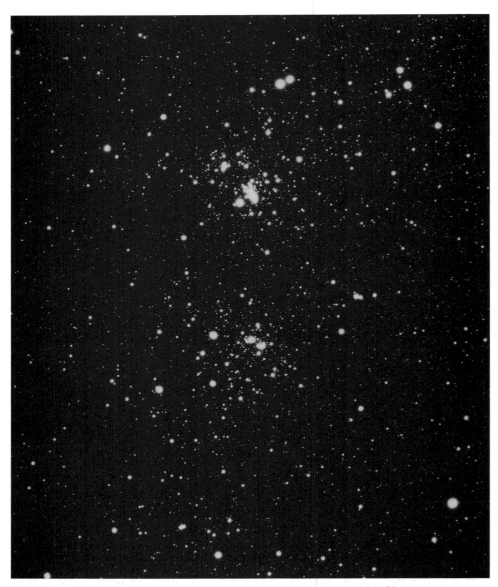

Fig. 13-4. The Double cluster, CHI & H Persei, in Perseus. (Courtesy Lick Observatory.)

ones can blend into the rich background of stars (Fig. 13-5).

If you have access to some of the photographic all-sky atlases like the *Palomar Sky Survey* or *Atlas Stellarum* this becomes less of a problem. Most of the clusters listed in the NGC can be picked out easily because these atlases have very deep limiting magnitudes. All you have to do is compare your field drawings with the atlas or trace the field from the atlas for comparison at the telescope. These atlases are very expensive, however, and out of the price range of the beginner. Vehrenberg's *Atlas Stellarum* is a two-volume set of charts that costs $225. Trust me on this, you don't want to know the cost of the *Palomar Sky Survey* plates.

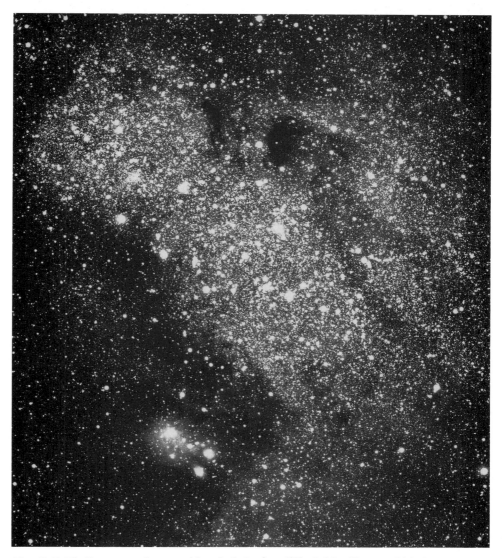

Fig. 13-5. Finding a star cluster amid such clutter is a difficult job. (Courtesy Lick Observatory.)

An easier and much less expensive way to get a peek at what your target looks like is to get *The Webb Society Deep Sky Observer's Handbook*. A seven-volume set, the Webb handbooks are indispensable references for anyone interested in deep-sky observing. The third volume of the series, *Open and Globular Clusters*, lists descriptions and drawings of 205 of the brightest open clusters in the sky. With the Webb Handbook, you can compare the drawing of the cluster directly with what you see at the eyepiece.

Open clusters are classified under a system devised by Robert Trumpler of the Lick Observatory in the late 1920s. His system rated open clusters according to three characteristics. The *concentration* of the open cluster can be:

I   Detached from the surrounding star field with a strong concentration of stars toward the cluster's center.
II   Detached from the surrounding star field with a weak concentration of stars toward its center.
III Detached from the surrounding star field with no apparent concentration of stars toward the cluster's center.
IV   Not well detached from the surrounding star field.

The second characteristic assigned by Trumpler concerns the *range of brightness* for the open cluster:

1   Cluster has a small range of brightness between its brightest and faintest member star.
2   Cluster has a moderate range of brightness between its brightest and faintest member star.
3   Cluster has a large range of brightness between its brightest and faintest member star.

The final characteristic is the *richness* of the cluster measured in the number of stars composing the cluster:

p   Indicates a poor cluster, having less than 50 stars.
m   Indicates a moderately rich cluster with between 50 and 100 member stars.
r   Indicates a rich cluster with over 100 stars.

A cluster like the Pleiades has the classification of I-3-r, indicating that it appears detached from the surrounding star field, has a large range of brightness from the faintest to brightest members, and has over 100 stars associated with it.

The second type of star cluster is the *globular cluster*. Globulars are compact groups of 100,000 or more stars squeezed into a distinctive spherical shape. These objects, because of the vast numbers of stars they contain, appear significantly brighter than open clusters and are much easier to pick out from the stellar background.

Like open clusters, globular clusters are classified using a scale denoting their concentration. This scale was devised by Harlow Shapley of the Harvard College Observatory in the 1920s. The classification of globular clusters is much simpler than that for open clusters. It ranges from 1 for the most highly concentrated, to 12 for the least concentrated cluster. The Great Hercules cluster (M13) rates a 5 on Shapley's scale, telling the observer that it is of about average concentration. The thing to remember when using this scale is that the smaller the number, the higher the concentration of stars in the object.

STAR CLUSTERS FOR THE TELESCOPE

Probably the best-known open cluster is the Pleiades (M45). It has been known and recognized as a related group of stars since the days of Homer. Easily found with the naked eye in the constellation Taurus, the cluster is best seen with a wide-field, low-power eyepiece or with binoculars. To the naked eye from 6 to 12 stars have been reported visible. With a small telescope, close to 80 stars are visible.

Fig. 13-6. The Pleiades (M45) in Taurus. (Courtesy Lick Observatory.)

During the last half of the nineteenth century, a debate raged concerning the visibility of a nebula surrounding the Pleiades member Merope (Fig. 13-6). Observers said that this nebulosity was easily visible in a small instrument, but when a larger one was turned on the object it was invisible. One observer said that the nebula was visible with his 2-inch finder, but invisible with the 11-inch main telescope. S. W. Barnard, using the 18-inch Alvan Clark refractor at Dearborn Observatory, could not see the nebula. The nebula does exist. In fact, the entire cluster is riddled with vaporous nebula. It is heartening to know that there are some things in the universe that are visible to the amateur with a small instrument and denied to the owners of larger telescopes!

Another naked-eye open cluster is found in the constellation Cancer. The Beehive cluster (M44) was seen by Galileo to contain ''a mass of more than 40 small stars.'' Because the cluster is so large, almost 2 arcminutes across, it is best seen in wide-field instruments like a rich-field telescope or even binoculars. The Beehive has a Trumpler classification of II-2-m.

A final naked-eye cluster is really a pair of open clusters aptly named the Double cluster. Easily found midway between the *W* of Cassiopeia and the main body of Perseus, they appear as a bright, hazy spot against the sky. For some reason these beautiful clusters were overlooked by Messier and are not included in his catalog. Officially known as NGC 884 and NGC 869, these objects make up one of the most beautiful sights for the small telescope.

For the best view, use an eyepiece with a field of at least one degree. The double cluster fills the field of view with a diamondlike sprinkling of stars. From their obvious intensity, the Trumpler classification of I-3-r for each cluster, just like the Pleiades, is well deserved. While you are drinking in the beauty of the clusters, take a few moments

Fig. 13-7. The open clusters in the constellation Auriga: M37 (a). (Courtesy Walter Piorkowski, ©1988.)

Fig. 13-7. The open clusters in the constellation Auriga: M36 (b). (Courtesy Walter Piorkowski, ©1988.)

Fig. 13-7. The open clusters in the constellation Auriga: M38 (c). (Courtesy Walter Piorkowski, © 1988.)

and see if you can detect colored stars in either one. There are a number of red super-giants in and around the clusters. The eastern cluster (NGC 884) has three red stars visible inside it and there is a very prominent red star between the two clusters.

Our next open cluster, M37, is only visible through a telescope. Located in the constellation Auriga at RA: 5h 52m Dec: +32° 33′, M37 is the brightest and best of the three Messier clusters in the constellation (Fig. 13-7). The biggest difference among the three Auriga clusters—M36, M37, and M38—is the brightness range of the stars in each cluster. All three are moderately rich to rich. Two, M36 and M37, appear to have a weak concentration of stars toward their centers. M38 has no apparent concentration. M37 appears the brightest because it has the smallest brightness range of the three combined with the richest membership, or more stars with less difference in their brightness.

M29 in the constellation Cygnus (Fig. 13-8) at RA: 20h 24m Dec: +38° 32′, on the other hand, is a difficult object to pick out of the background stars because it has no

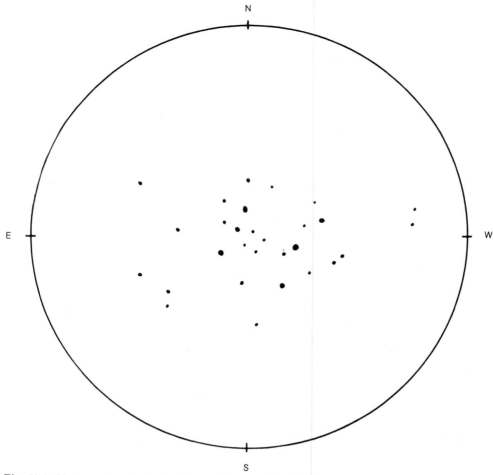

Fig. 13-8. M29, an open cluster in Cygnus. (Courtesy Dan Troiani.)

apparent concentration of stars at its center (III), has a large brightness range (3), and is a star-poor cluster (p). When I first looked for M29 with my 80mm I passed right over it, not once but three times.

NGC 2362 (Fig. 13-9), located in Canis Major at RA: 7h 18m Dec: 24° 57′ and centered near the magnitude 4 star Tau, is an example of what a strong concentration of a few stars can do to the overall appearance of a cluster. The cluster has a Trumpler classification of I-3-p. You would expect that, because it has a large brightness range and relatively few stars, so it appears as a dull little cluster. But with most of the stars concentrated toward the center of the cluster, it appears ''small but densely packed with faint stars'' in a 4-inch telescope.

Probably the best-known globular cluster is M13, the Great Hercules cluster. We have already had a peek at this with our binoculars, so let's compare it with our view at the telescope. With an integrated magnitude of 5.8, M13 is just visible to the naked eye when the skies are extremely transparent. The cluster was discovered in 1774 on a chance observation by Edmond Halley of Halley's comet fame. Messier observed it a short time later and declared emphatically that the nebula contained no stars. This object can be completely resolved by a 6-inch telescope at 130×, so we know Messier was wrong about it not containing any stars!

In 10×50 binoculars the cluster was visible only as a bright smudge against the background stars, but with a small telescope it begins to take on the classic characteristics of a globular cluster, with a milky center surrounded by glittering specks of starlight. One nice thing about globulars is that they take magnification fairly well, so don't be afraid to use a higher power. At 36× my 80mm shows a round shape with a very bright center. When I use 100×, I can resolve some of the outlying stars, but nothing near the center.

Located north of M13 in the same constellation at RA: 17h 17m Dec: +43° 08′, is the globular M92 (Fig. 3-8). With a Shapley classification of 4, M92 is slightly more concentrated than M13. Using a 4-inch SCT, Barbara Lux of McKeesport, Pennsylvania,

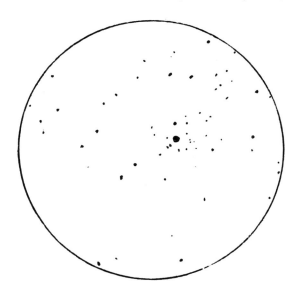

Fig. 13-9. NGC 2362, an open cluster in Canis Major. (Courtesy Suresh Sreenivasan.)

describes it as "bright and beautiful with its central condensation and diffused edges plainly seen."

Slightly less concentrated than M13 is M3 in the constellation Canes Venatici at RA: 13h 42m Dec: +28° 23'. With a Shapley classification of 6, M3 is seen as a looser cluster than either M13 or M92. A 6-inch telescope will resolve the stars around the edge of the cluster, but nothing toward the center of the globular.

Our final globular is the loosest of all we have seen. M4 can be found easily very near the star Antares in the constellation Scorpio. With a Shapley rating of 9, it is the only globular that we have looked at that can be at least partially resolved with a small telescope. A 3- or 4-inch telescope easily shows the outer edges broken into individual stars. The 4-inch will partially resolve the center of the cluster.

You might have noticed that many of the open-star clusters are found in the autumn or winter sky, while the globulars are found in the spring and summer sky. There is a simple reason for this. It relates to how the clusters are located throughout the galaxy. Open clusters form in the spiral arms of the Milky Way galaxy, while globular clusters tend to lie around the center of the galaxy. Many, but not all, globulars are found in a halo that orbits like satellites around the center of the galaxy. When we face outward away from the center of the galaxy, as we do during the winter months, we are looking through the arms of the galaxy. As a consequence we see many more open clusters than when the Earth's orbit puts us in line to look toward the center of the galaxy, as happens during the summer months when we can see more globular clusters.

A number of other clusters suitable for observation with a small telescope are listed in Appendix D.

NEBULAE

At one time the term *nebula* referred to a number of faint, apparently glowing objects including galaxies, which are completely different as we will see. Today, astronomers have broken nebulae into three major classes.

Dark nebulae are lanes of dust thick enough to block the light from the stars behind them. They appear as dark lanes seemingly void of stars, and can be found along our line of sight toward the center of the galaxy. A dark nebula can block light coming through it by as much as three magnitudes. The Great Rift that runs from Cygnus to Sagittarius is a good example of a dark nebula. The Milky Way seems to be split in two as it runs south out of Cygnus, through Aquila, and into Sagittarius where it recombines into one large star field. The Great Rift is caused by dust lying between us and the center of the galaxy, the direction in which we look when facing those constellations. Dark nebulae are best observed with nothing more than the naked eye.

Bright nebulae are knots of gas and dust that are illuminated by stars buried within or around the nebula. Bright nebulae can be illuminated from behind forming *emission nebulae*, type E, which shine from the light of stars embedded in them and whose light passes through them. Bright nebulae also can shine because of light from nearby stars reflecting off the dust that makes up the nebula, forming a *reflection nebula*, type R. Bright nebula also can be made by a combination of these two light sources, forming an *emission-reflection nebula*, type E + R.

Planetary nebulae are the shells of stars blown off into space as the star ages. The expanding shell of matter is illuminated by the light of the star that ejected it. This star is found most often near the center of the nebula, but is usually too faint to be seen in most amateur telescopes. The similarity in shape to a planet and the nebula's regular outline led Sir William Herschel and other eighteenth century observers to call them ''planetlike'' and the name stuck.

Planetary nebula range from starlike, stellar points of light to rather irregular forms. Small planetaries, particularly the stellar types, can be difficult to pick out from the background stars. You can help identify these planetaries by using a diffraction grating. Hold it between your eye and the eyepiece and move it quickly back and forth. Because stars shine by self-produced light, their images will turn into tiny streaks showing their spectra. But because the planetary nebula shines from second-hand light from its star, it will not streak and appears as a tiny point of light. When using this technique, the planetary will appear the same with and without the grating.

NEBULAE FOR THE TELESCOPE

The most well known of all bright nebulae is found in the constellation Orion embedded near the tip of that figure's sword (Fig. 13-10). M42, the Orion nebula, is the classic

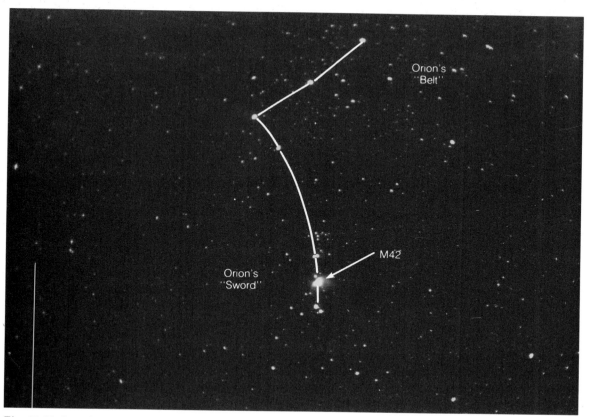

Fig. 13-10. Location of M42 in Orion. (Courtesy Lee Keith.)

Fig. 13-11. The Great nebula of Orion, M42. (Courtesy Lee Keith.)

example of the combination emission-reflection nebula (Fig. 13-11).

Observers get a bonus when observing this object. The nebula we see is part of a vast complex of nebula and stars. In addition to the bright emission-reflection nebula that we see is a dark nebula in the complex called the Fish's Mouth, which is comprised of dust and gas lying between us and the Orion nebula. Near this dark nebula, embedded in the wisps of gas, is a group of stars known as the Trapezium, which have burned their way through the bright obscuring dust that makes up the E + R nebula.

Although the Orion nebula is considered by many to be the most spectacular object on Messier's list, it is interesting to note that Galileo never turned his instrument on it. The discovery of the Orion nebula had to wait 46 years until 1656 when Christian Huygens turned his telescope on the stars that make up Orion's sword and made a double discovery. First he found the Great Orion nebula and three stars close together. The fourth, first seen by Hooke in 1664 but not recorded, was rediscovered by Struve in 1826. The stars of the Trapezium and the Orion nebula are linked together because the nebula is the birthplace of the stars, a stellar nursery.

Also located in Orion is a fine example of a reflection nebula, M78. Located above the belt of Orion at RA: 5h 47m Dec: +0° 03', M78 is easily seen even under moderately dark skies. Using an 8-inch SCT from his home just south of Chicago, Suresh Sreenivasan describes it as being a "light wispy nebulosity with two bright clumps of nebulosity embedded in it." (Fig. 13-12).

One of the easiest planetary nebulae for the small telescope is M57, the Ring nebula (Fig. 13-13). It was first observed by Messier early in 1779. He described it as looking like a "patch of light, rounded but unresolvable" when viewed by the best telescopes of

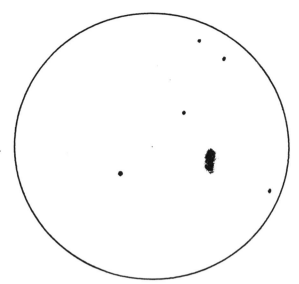

Fig. 13-12. The nebula M78 in Orion.
(Courtesy Suresh Sreenivasan.)

the time. Decades later, Sir John Herschel said it looked like a piece of gauze stretched over a hoop.

For most small instruments this is an easy object, despite its lack of surface brightness and low apparent magnitude at magnitude 9. This is because the Ring nebula sits almost alone in its field, with no real distractions to divert the eye's attention. With an

Fig. 13-13. The Ring nebula, M57, in Lyra. (Courtesy U.S. Naval Observatory.)

80mm refractor the Ring nebula is easily found, even against the light-dappled skies over Chicago. A dim, almost grayish-looking object, it takes magnification well.

Located in the constellation Gemini at RA: 7h 29m Dec: +20° 55′ is a real challenge to your growing observation skills. NGC 2392, the Eskimo nebula (Fig. 13-14), is a challenge to find for the small-scope user. It challenges the owner of a larger instrument to pick out the details that gave it its name. You will know this planetary when you find it by its blue-green color. I recommend using the grating technique described earlier to help you along. If you are using a scope less than 6 inches in aperture, this faint, magnitude 9.9 object may reveal the outer areas of its shell, but you will have to use averted vision. The 8-inch telescope will begin to reveal the mottling around the edges of the object, which makes it look like a tiny face peeking out from under a parka. Again I warn you that this is a very challenging object. Do you want to try it?

Other nebulae suitable for observation with a small telescope are listed in Appendix D.

GALAXIES

Galaxies are probably the most challenging of all deep-sky objects. The vast numbers and wide diversity of these "island universes" that inhabit the night sky make them common targets for deep-sky telescopes.

For many of these objects, however, the telescopic requirements are a bit stiffer than you have experienced so far. Most galaxies are not only faint, but they have very low surface brightnesses, which makes them impossible to resolve in most amateur instruments. But galaxy hunting is a fun side road for the amateur to venture to from time to time. If you have a telescope in the 8- to 10-inch range you might even make a significant contribution to science by discovering a supernova.

The key to observing a galaxy is the orientation it takes toward Earth. If the galaxy is almost face on, like M83 in Fig. 13-15, it has better contrast with the background and is easily seen. A galaxy turned on edge, like NGC 4565 in Fig. 13-16, offers poorer contrast and is more difficult to see.

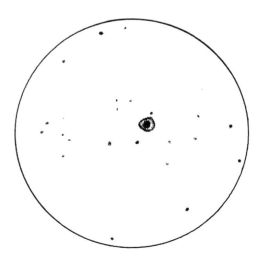

Fig. 13-14. NGC 2392, the Eskimo nebula, is a planetary nebula in Gemini. (Courtesy Suresh Sreenivasan.)

Galaxies for the Telescope

At 2.2 million light years, the Andromeda galaxy (M31) is at the very limit of the universe that can be seen with the unaided eye. There are many objects farther away, but none that appear as majestic when viewed from our little corner of the universe. To the unaided eye, the Andromeda galaxy appears to be a wisp of smoke or cloud lost among the stars. Over 1,000 years ago, the Persian astronomer Al-Sufi described this galaxy as "a little cloud" in his *Book of the Fixed Stars*.

The first person to look at the galaxy with a telescope was Simon Marius. A contemporary of Galileo, Marius turned his simple telescope on the Great Andromeda nebula in December 1612. He concurred with the cloudlike nature of the object, but noted that it "consisted of three rays; whitish, irregular and faint; brighter towards the center." Edmond Halley, second Astronomer Royal, studied it also and said it appeared to emit a beam of light. Charles Messier cataloged it on August 3, 1764, as entry 31 in his list of objects a serious observer should not confuse with comets.

M31 is easily found with the small telescope. Located just above Nu Andromeda, it is easily visible as a greenish glow with a low-power, wide-field eyepiece. It takes a long, time-exposure photograph to reveal the stars that make up M31, but you will be able to trace the extent of the galaxy from its luminous central area to the points where it fades out to the east and west of center.

On a night with excellent transparency you should be able to detect M31's compan-

Fig. 13-15. The spiral galaxy M83 in Hydra. (Courtesy Dan Troiani.)

ion galaxy just south of the galaxy's nucleus (Fig. 13-17). This smaller galaxy, M32, shines at magnitude 9. Also visible to the small telescope is another companion, M110. Located to the northeast of the nucleus, this galaxy is slightly dimmer at magnitude 10.8, but larger than M32.

Located in the constellation Canes Venatici at RA: 13h 29m Dec: +47° 12′, M51 is really two objects joined together. The primary object is M51 at magnitude 8.4. It is a face-on spiral discovered by Messier in October 1773. At the tip of one of the spiral arms lies NGC 5195, a companion galaxy discovered by Mechain in 1781.

Although Messier never saw the companion, it shows faintly in most small telescopes available today. So consider the fact that even today's smallest instruments are superior to those available to some of astronomy's greats! A small telescope will show the galaxy as a hazy spot with a bright nucleus. The spiral structure becomes more apparent with larger instruments.

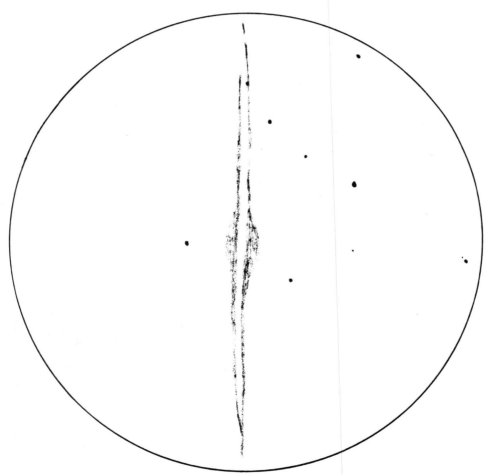

Fig. 13-16. The edge on galaxy NGC 4565 in Coma Berenices. (Courtesy Dan Troiani.)

Another spiral galaxy like M31 is found in the constellation Ursa Major at RA: 9h 56m Dec: +69° 04'. M81 is magnitude 6.9 and similar to M31 in type. It is turned so that it almost faces us. Discovered by Frederick Bode in December 1774, it was described as a "nebulous patch, more or less round with a dense nucleus in the middle." When Messier added this object to his catalog in February 1781, he described it as a bright little oval. Many of the observers of M81 are impressed with the brightness of the object. Rev. T. W. Webb, writing during the mid-nineteenth century, said M81 is "bright with a vivid nucleus."

This is another object where the observer is treated to a two-for-one sight. In the same field is another galaxy, M82. Classed as an irregular galaxy, M82 seems to be undergoing the throes of an explosion beyond imagination.

Other galaxies suitable for observation with a small telescope are listed in Appendix D.

GALAXIES, SUPERNOVAE, AND THE AMATEUR

The study of galaxies is pretty much in the hands of professional astronomers and their big telescopes and sophisticated equipment. There is one area in which the amateur can

Fig. 13-17. The great Andromeda galaxy, M31, and its smaller companion M32. (Courtesy Dan Troiani.)

make a contribution, however. That is the systematic monitoring of galaxies for supernovae. The requirements for this project do include a telescope over 8 inches in aperture, but with the explosive growth of large-aperture amateur instruments many amateurs already have the tools they need.

A supernova is the massive explosion of a star accompanied by its rapid, dramatic increase in brightness. Stars in another galaxy are not usually visible in the amateur telescope, but a supernova can be seen as a point of light that brightens very rapidly, usually in one or two days, and then decreases in brightness over a relatively short period of time. The trick in maintaining such a program is knowing what the galaxy you are watching looks like without any extra points of light.

This is not as hard as it may seem. First off, the probability that a galaxy you are monitoring for the first time will have a supernova in it is a bit rare. These things don't go off every day! A supernova is a very rare event—only two have been recorded in our galaxy in almost 1,000 years. Rev. Robert Evans of Australia is the leader in visual supernova discoveries. Since 1981 he has averaged two discoveries per year. When you factor in the total number of observations he makes each year, this comes to about 1 discovery for every 10,000 observations! And they say comet hunting is tough!

If you can find a copy of *Hubble's Atlas of Galaxies*, by all means grab it! Not only is it a beautiful book, but it also shows the appearance of many galaxies in their normal, non-supernova state. All you have to do is become familiar with a galaxy, something you will do by repeated observation anyway, and you will notice if something goes amiss out there.

It is best to limit your monitoring to galaxies that show you a large portion of their face, usually spiral galaxies. It is in these types that you have the best chance of success. The Messier catalog can give you about 26 potential candidates for a systematic supernova patrol. They are all spiral galaxies and have an integrated magnitude brighter than 10. The entire procedure is simple. Check each galaxy once every clear night. If you see something contact the Smithsonian Central Bureau of Astronomical Telegrams as soon as you are sure of your discovery.

DRAWING THE DEEP SKY

I know we've talked about this before, but drawing deep-sky objects is different from drawing the Moon and planets. No two deep-sky objects are alike, even if they are both of the same type. You can record descriptions of the objects in your log, but take a good look at M42 and really try to describe it. Pretty hard to do isn't it? It's not that written descriptions aren't important. They are. They allow you to improve your observational skills and look at objects with a critical, analytic eye. But if written descriptions can do that, a drawing of a deep-sky object will help develop the eye even faster.

Unlike planetary drawing, there are no preprinted forms you can use to sketch a deep-sky object. All you need is a plain piece of white paper with a circle on it. The circle is to keep you honest, so to speak, and indicates the boundaries of your field of view. The size of the circle is up to you. Some observers use a 2-inch circle; some, a 3-inch circle. Use whatever size makes you comfortable. Some nice sharp pencils—the kind used in drafting with different hardness leads—are good, an eraser, and a stick of graphite round out all you need to start sketching.

Get comfortable at your scope and look at the object you want to draw with a couple of different eyepieces. Use the eyepiece that shows you the most detail for your sketching. Begin by filling in the background stars around the object. Be careful about the placement of the brighter stars in the field; work from the center out to the edges of the field. Next add any little asterisms that catch your eye. Including them can be helpful in making a quick identification of an object you don't know.

If you are drawing an open cluster, carefully place each star in its appropriate place. Open clusters are tough because there are usually so many stars to record so they can take a while to draw. If you are drawing an object with a nebulosity, rub some of the graphite onto the paper and then use your thumb to smear it around into the approximate shape of the object. The first smear will give you a gray tone. If you want to darken it, wet your thumb and rub it again until you have the desired tone. To lighten it, use your eraser to skim off a bit. If you are recording the appearance of a globular cluster, rub on some graphite before you add the stars that are visible in your telescope. Then add the stars that you can resolve. This gives the globular the same 3-D effect it has in the eyepiece.

A planetary nebula like the Ring nebula is easy to draw. It has a definite boundary that you can outline easily. If you detect shadings in the nebula, add them by using the flat or the point of the graphite stick. Add the central star only if you can see it.

Your drawing should be an accurate representation of what you see in the eyepiece. Don't add anything because you know about where it should be on the drawing. Galaxies, like planetaries, show a definite outline, so draw that first and then work in toward the nucleus. If you are cursed with a large telescope, you also will have to add details of your view like dust lanes and bright patches.

Bright nebulae will give you the most trouble. With their wide range of brightness and subtle shadings across their surface, it is often best to smear down a uniform layer of graphite and make it darker or lighter as you move out from the center of the object.

When you are finished with your sketch, you will have an important addition to your hopefully growing observation log.

14

Observing
The Sun

*The influence of this eminent body . . . is so great . . . that it
becomes almost a duty to study the operations which are
carried on upon the solar surface.*

Sir William Herschel

Sometimes I do not understand astrophysicists. They spend their nights trying to fathom the life cycle of a small, average star. In the process they stay up long hours and rob themselves of sleep when all they have to do is roll out of bed, open their shade, and look at the greatest stellar laboratory imaginable.

The Sun is a special object for the amateur astronomer (Fig. 14-1). It must be treated with respect and dealt with carefully. It is the only astronomical object that can physically harm you. Galileo learned the hard way. Turning his telescope on the dim image of the Sun at sunset he was shocked to learn that the intensity of the Sun wasn't really diminished. After that first look he was blind for a week. Even after his sight returned, the aftereffects followed him for the remainder of his life in the form of deteriorating eyesight. Because he took one short peek at the unshielded Sun, Galileo died blind.

*Never look directly at the Sun through a telescope or binoculars
of any kind without proper protection.*

Do you remember those novelty lighters that light the cigarette by using a mirror to focus the rays of the Sun on the cigarette's point? Well, a binocular or telescope objective does the job infinitely better. The intensity of the light focused by even a small objective lens or mirror produces sufficient heat to set a solid block of wood on fire! You do not want to put your eye anywhere near that kind of heat.

TELESCOPES, FILTERS, AND THE SUN

Most telescopes purchased at the local department store come with a small, screw-on type of filter that supposedly reduces the intensity of the Sun's image. If your telescope came with one of these filters, the best way to use it is to grasp it firmly in your hand and *throw it as far away from you as possible*! These so-called filters *are not safe*. Remember, the optical system of a small telescope, even the finder, will produce an intense amount of heat. If you put this little filter at the point of focus of that heat, it is going to crack. Even a tiny crack will allow enough of that heat into your eye to cause permanent damage.

There are only two safe ways to observe the Sun. First, you can reduce the intensity of the solar image *before* it passes through the telescope's optical train. Reflective aper-

Fig. 14-1. The Sun. (Courtesy Walter Piorkowski, ©1988.)

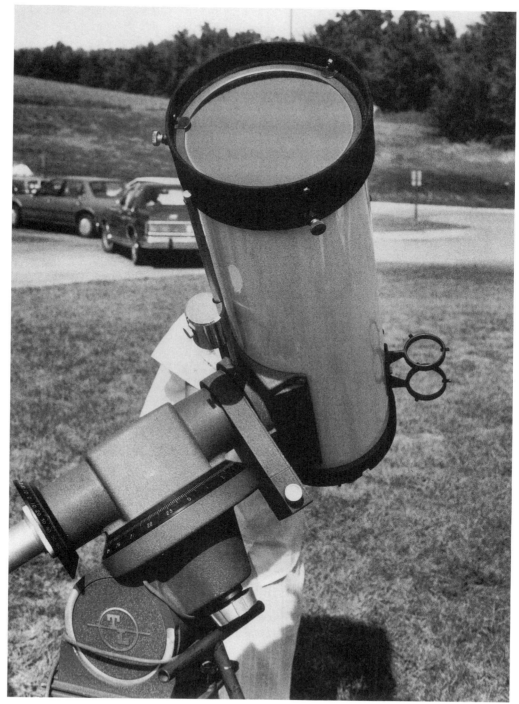

Fig. 14-2. Full-aperture Mylar solar filter on an 8-inch cassegrain telescope.

ture filters are made of a thin layer of metal, usually Inconel, that is evaporated over an element of optical glass. This in turn is aluminized on both the inside and outside layer so that the light from the Sun is reflected *away* from the telescope. It allows only about .1% of the Sun's normal brightness to enter the optical train. By reflecting the brightness of the Sun away from the telescope, the heat caused by the intense solar image is prevented from entering your telescope's optical system. Using an aperture filter is the *only* way for completely safe, direct viewing of the Sun.

A full-aperture filter (Fig. 14-2) is the same size as your objective. It completely covers the front end of your telescope and looks like a mirrored end cap. A subdiameter aperture filter (Fig. 14-3) also fits over the end of your instrument, but it is mounted with an aperture mask like that used for planetary observing, and the filter surface is smaller. Full-aperture filters are fine for small refractors and instruments with an unobstructed light path. Subdiameter filters work best with reflectors and catadioptrics with a large secondary obstruction.

Aperture filters are made by a number of companies (see Appendix I for a complete list). They can be made with two types of filter material, each with different transmission ranges. Glass filters made with Inconel produce a yellow image of the Sun. Filters can also be made with a thin sheet of Mylar, which gives the Sun a blue image. Because the eye is less sensitive to the blue end of the spectrum, the Mylar filters tend to lose small details on the solar surface. I use a full aperture glass filter made by Thousand Oaks Optical Company (Fig. 14-4) on my 80mm refractor. It gives a yellow image and I'm very pleased with the amount of detail it allows me to see on the Sun.

A subdiameter filter is preferred over full diameter filters in some instances. The same considerations given to atmospheric conditions affecting seeing at night apply to observing the Sun during the day. The same air cells that cause erratic seeing when you observe the planets and the Moon can affect your observations of the Sun. This is further complicated during the day because the Sun is heating the air, which rises as "thermals" off of everything: roofs, pavement, garden sheds, everything. When you use the subdiameter aperture filter you are better able to look through less of these cells so the image improves.

If you use a reflective aperture filter of any kind, there is one caution that bears passing along. *Cover your finder scope or get an additional filter for the finder.* The finder, after all, is a telescope in its own right and will act to intensify the heat from the Sun. Although you may not be foolish enough to look through it, someone else may not have your common sense. I hear a couple of you asking a logical question. "How are you supposed to find the Sun if your finder is covered?" Well, it's really quite simple. Just look at the shadow of your telescope (Figs. 14-5). Move the telescope around until you get the smallest shadow, then you know the scope is pointed at the Sun.

> *Never look directly at the Sun through a telescope, finder scope,*
> *or binoculars of any kind without proper protection.*

The other safe method of solar observation is the *projection method*. This is both the easiest and safest method for observing the Sun. (Fig. 14-6). With the projection method an image of the Sun is passed through your telescope and projected onto a flat

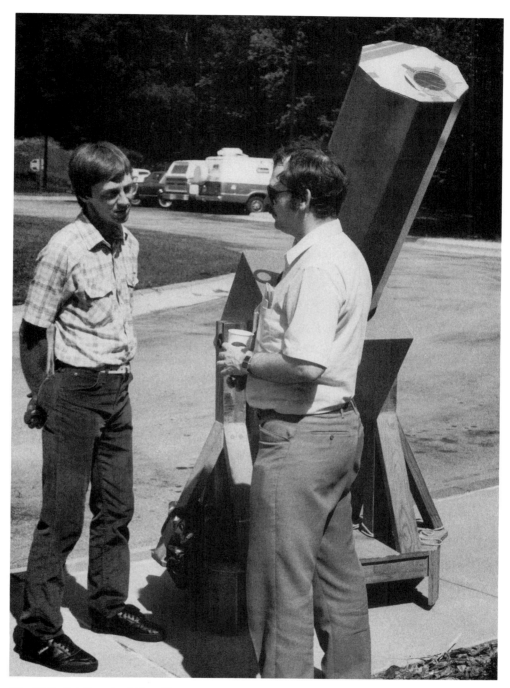

Fig. 14-3. Subdiameter Mylar filter on a 10-inch reflector.

white surface, like a sheet of paper or a card. You observe this image by facing the projected image with your back to the telescope and its eyepiece.

One thing to remember when using the projection method is that the Sun is unfiltered, and therefore undiminished in intensity, as it passes through your telescope. This limits the type of eyepiece that you can use to project the solar image. Any eyepiece that contains multiple elements cemented together, like a plossl, orthoscopic, or RKE, cannot be used. They are going to be placed where the solar image comes to a focus, right at the point of the image's greatest intensity. Because of this the cement that binds the elements together will be subjected to intense heat and will melt, rendering the eyepiece useless. It is better to use an eyepiece with single, noncemented elements like a rams-

Fig. 14-4. A full-aperture reflective filter on an 80mm refractor. Note the finderscope is covered.

den or huygens. This is why I told you in chapter 5 to tuck them away in a drawer. SCTs should be used with an aperture mask even when using the projection method because the heat will build up inside the sealed tube and possibly damage the cement holding the secondary to its mount.

The actual construction of a projection system is easy. First determine the distance between the projection screen and the eyepiece. Do this by holding a white sheet of paper about 12 inches away from the eyepiece and carefully focusing the image of the Sun on the paper. The image should fill the paper and allow you to easily see any visible details. If it's not sharp enough, move the paper back and forth until you get a sharp, detailed image, then measure that distance. Now use a thin strip of wood or metal (metal bends easier) to make the support for the projection screen. Attach it to the telescope with any convenient type of clamp. Large hose clamps work nicely for this.

Fig. 14-5. To find the Sun aim the fully filtered (with finder covered) telescope in the general direction of the Sun and watch the shadow. When the scope is far away from the Sun, the shadow will be long (a). As you point the telescope closer toward the Sun the shadow becomes shorter (b), until it is smallest when pointed directly at the Sun (c).

Because the image of the Sun grows dimmer as it moves away from the eyepiece and onto the screen, it can be overwhelmed by the glare of the Sun itself. If the optical train of the telescope and eyepiece form a straight line, use a second piece of paper or posterboard to shade the screen (Fig. 14-7). Even with a newtonian it is not a bad idea to shade the screen with an oversized piece of posterboard. Bend it so that it acts as its own shade.

Both the reflective aperture filter and the projection method allow you to observe the Sun in normal white light. Observations can also be made in *monochromatic light* using special devices that filter the Sun's light to admit only one wavelength. This wavelength is usually the *hydrogen-alpha band* and it can afford some absolutely spectacular views of the Sun that are missed with white light. Equipment to see this band of light, in the form of hydrogen-alpha filter systems, are expensive. So we will confine our look at the Sun to those features you can see in white light.

THE SOLAR LANDSCAPE

Okay, reflective aperture filter or projection screen in place? Remember,

Never look directly at the Sun through a telescope or binoculars
of any kind without proper protection.

The Sun looks peaceful enough, but looks can be deceiving. Solar activity takes place in 11-year periods called *solar cycles*. During periods of *solar minimum*, the lowest phase of activity, the Sun can seem a pretty dull place. But for every minimum there is a

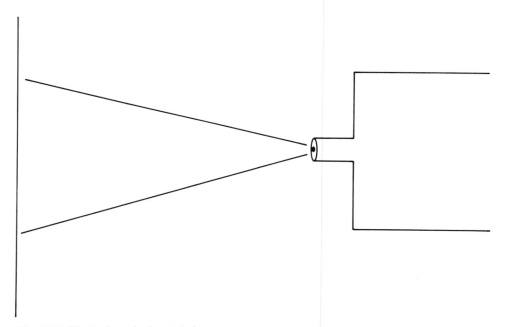

Fig. 14-6. The basic projection technique.

corresponding *solar maximum* during which the sun really heats up. Throughout the solar cycle there are a number of features that the amateur can observe and record easily.

At first glance the bright solar surface, or *photosphere*, seems flat and featureless. But if you watch and the seeing is excellent, you will notice that the seemingly featureless photosphere is covered with tiny splotches, almost like the graininess of a photograph when viewed up close. This graininess is caused by small cells called *granules*. Granules range in size from 1 to 5 arcseconds across with an average size of 2.5 arcseconds.

Granule visibility is the measure from which solar observers can determine seeing quality. When the seeing is excellent, better than 1 arcsecond, the granules are easily resolved and seen. If the granules are seen as mottled or spotty, the seeing is rated at 1 to 2 arcseconds or good. If the granules appear to pop in and out of view, the seeing is 2 to 5 arcseconds and rated as fair. Anything beyond this 5 arcsecond limit is poor seeing.

The most obvious feature of the solar disk are the dark *sunspots*. Sunspots are relatively cool features that float on the photosphere. If you look carefully, you will see that a sunspot is more than just a spot (Fig. 14-8). The dark, central area is called the *umbra*. It is usually, but not always, surrounded by a lighter, but still dark, *penumbra*. The penumbra itself can show a lot of detail. They can be seen to contain, if the seeing is good, dark areas called *filaments* and brighter areas called *grains*. These may be overlayed with a system of *fibrils*, thin dark lines that radiate outward from the umbra. Very small sunspots that have no penumbra are called *pores*.

Sunspots tend to form in pairs and cluster in magnetically related groups as they move across the photosphere. Sunspots and sunspot groups come in a wide variety of shapes and sizes. To simplify and standardize descriptions of these groups the *Zurich sunspot classification system* was devised. Sunspot activity follows the 11-year solar cycle. At solar minimum there may be few, if any, sunspots, while at solar maximum the surface can appear riddled with them. The last period of solar minimum was 1986; 1990

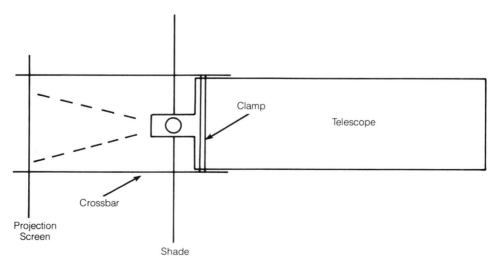

Fig. 14-7. A projection screen should be shaded from the glare of the Sun.

is the predicted period of maximum solar activity. Sunspots rotate with the Sun, completing a period in a little over 27 days.

Seen near sunspots are bright patches called *faculae*. Because they are so bright, faculae have little contrast against the almost equally bright background of the photosphere. The Sun has an atmosphere, called the *corona*. When we look at the center of the solar disk we are looking down, straight through the Sun's atmosphere. When we look at the limb of the Sun we are looking through more of the corona so the limb looks darker than the rest of the solar disk. This *limb darkening* gives you your best chance to see the faculae because it will have much more contrast against the darker limb than the brighter regions away from the limb.

Another type of bright patch, which is brighter than the umbra of the sunspot, can sometimes be seen crossing the sunspot's umbra. These are called *light bridges*. The appearance of a light bridge in a sunspot usually indicates that the spot is about to break up and dissolve into the photosphere.

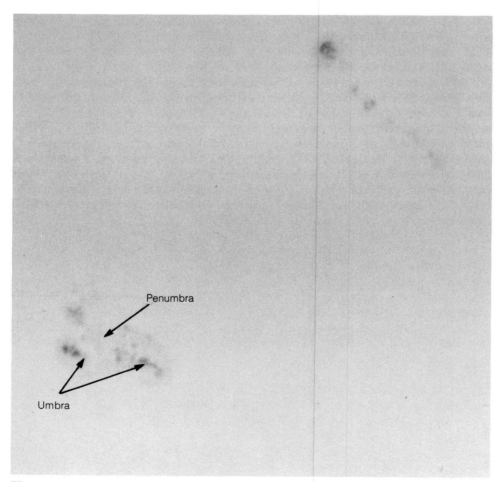

Fig. 14-8. Sunspots and sunspot groups. (Courtesy Walter Piorkowski, ©1988.)

Finally there is the ultimate bright spot that is seen near a well developed group of sunspots. *White light flares*, a very rare phenomenon with only about 12 recorded since the invention of the telescope, can appear as a sudden brightening in or near the sunspot group. They are so bright that they stand out clearly from the bright photosphere, but they only last a brief time.

CONTRIBUTIONS FROM THE AMATEUR

Accurate counts of sunspots are a reliable and important gauge of where the Sun is during the solar cycle. The AAVSO Solar Division collects and correlates sunspot counts from amateurs around the United States and the world. Counting sunspots is an easy and scientifically necessary activity that any amateur can do. To count sunspots, set up your telescope with its filtering or projection equipment and use an eyepiece that allows you to see the entire solar image. Then count the groups that you see.

Groups are usually quite obvious. If you see one, lonely little sunspot off by itself nowhere near the other groups on the photosphere, it also must be counted as a group. Don't trust your memory on this number, write the final total down in your log immediately after counting the groups. Next, switch to a higher magnification and scan the photosphere for any small spots not associated with the groups. Again, record this number in your log. Finally, return to each group and carefully count the number of spots that each contains.

When your observation is complete you will have three sets of numbers: the number of groups, the number of single, isolated spots; and the number of spots in the groups. To determine your own daily count, or the day's *relative spot number* (RSN), multiply the number of groups by 10 and then add the number of spots. Remember to count each isolated sunspot as *both* a group and a spot. For example, if you count 5 groups of spots, 2 isolated spots, and 13 spots within the 5 groups, your RSN would be 85. This is determined by adding 5 groups plus 2 isolated spots to get 7 groups ($7 \times 10 = 70$); then add 13 group spots plus 2 isolated spots for 15 spots: $70 + 15 = 85$. Quite an active Sun!

It is best to try running a count for awhile to determine your accuracy. Each month *Sky & Telescope* magazine publishes sunspot numbers for previous months compiled by AAVSO and by the Sunspot Index Center in Brussels, Belgium. Check your relative sunspot numbers against these in order to determine your accuracy. If you are close to the published figures, you should contact AAVSO to request information about joining the Solar Division. You can also pass your early data along to AAVSO for them to analyze.

While the AAVSO Solar Division concentrates on obtaining accurate relative sunspot numbers, the ALPO Solar Section 'ALPOSS' is concerned with the morphology, or the form and structure, of solar phenomena. In order to better study and analyze these events, ALPOSS needs to compile a daily visual record of the solar disk. Yes, I know this means drawing something, but solar drawings are among the easiest astronomical drawings to make. ALPOSS wants daily, full-disk drawings of the Sun.

To make these you first have to determine the inclination of the solar axis relative to Earth. Figure 14-9 shows this solar "wobble" through the course of a year. An accurate representation of the location of the poles is necessary for an accurate drawing. In your log book draw a circle to represent the Sun. Mark off the poles with "ticks" and designate north and south for orientation. Using an eyepiece that allows you to see the entire

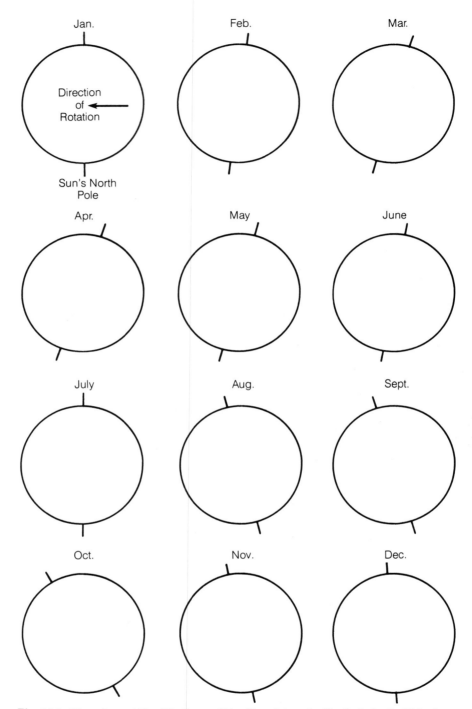

Fig. 14-9. The solar wobble. The Sun wobbles like a top as the Earth circles it. This shows the location of the Sun's north and south pole throughout the year.

solar disk, mark the positions of each spot or group on the disk. Use dots and dashes to indicate the outlines of the features. Be careful to accurately draw the placement, shape, and size with relation to the other groups on the photosphere and the solar disk itself.

Now go back and fill in the details in the individual groups. Darken the areas to indicate a spot's umbra, again paying attention to its proportion with relation to the other spots in the group. Outline the boundary of the penumbra, which need not be shaded. Solar projection observers have a bit of an edge with whole disk drawings. A sheet of paper can be placed on the projection screen and the detail traced directly off the projected image. The final drawing can be transferred to the back of the ALPOSS report form (Fig. 14-10).

You can also include detailed drawings of individual spots and groups. First, draw an accurate outline of the umbra region and shade it. Then outline the penumbra and shade it with an eye toward the varying intensities of the penumbra using light shading for light areas and darker shading for darker areas (but not as dark as the umbra). Be sure to represent the more apparent filaments and bright knots you may see in the penumbra. Also be careful in your depiction of the size of the group with relation to the background you are using. Most observers tend to make their regions too large. Again, projection users have a distinct advantage because they are working from a projected image that can be traced easily.

SOLAR ECLIPSES

One of the most awe-inspiring sights in nature is a total solar eclipse (Fig. 14-11). The slow darkening of the landscape caused by the advance of the Moon across the face of the Sun is eerie. The abrupt cutoff of light at totality brings features invisible during normal observing into view. The ghostly, shimmering corona is visible, as are the red *prominences*—huge eruptions extending hundreds of thousands of miles into space. Totality, however, doesn't last very long, at the most about seven minutes.

Unfortunately, total solar eclipses are rare events and are confined to very narrow strips on the Earth's surface. Predictions of solar eclipses for the year can be found in the *Astronomical Almanac* or the RCAS's *Observer's Handbook*.

More common are eclipses that occur when the Moon's shadow does not quite reach the Earth's surface. During these *annular eclipses*, the Moon is seen to be outlined by the bright disk of the Sun, which it doesn't quite cover (Fig. 14-12).

During a total eclipse the amateur can do a few things that can't normally be done. First, you can search for comets almost right up to the now dark solar disk. Because totality is so short, comet searching is best done with a camera. A series of exposures may net a comet as it rounds the Sun at perihelion. You can also easily find elusive Mercury and Venus. If they are not too close to the Sun you can keep them in telescopic view as totality ends, then observe them against the brightening background of the sky.

A.L.P.O. Solar Section
Active Region
Drawing
Report From

1. Date/Time _____ UT

2. Rotation No. _____, S.E.S.C. Region No. _____

3. Central Meridian _____, Hour Angle _____

4. Relative Heliographic Coordinates _____

5. Sky Conditions _____, Seeing _____, Clouds _____, Wind_____

6. Observatory: _____roll off roof, _____ roll off shed, _____dome, _____none

7. Aperture _____, Focal Length _____,f/

8. Instrument Type _____, Mounting Type_____

Clock Driven? _____

9. Eyepiece Type _____, Focal Length _____, Barlow? _____

10. Magnification _____, Drawing Scale _____

11. Observers File No._____

12. Observers Address and Telephone No.

Drawing Below

_____

Fig. 14-10. ALPO Solar Section report form.

Fig. 14-11. A total eclipse of the Sun on February 26, 1979. (Courtesy Walter Piorkowski, ©1979.)

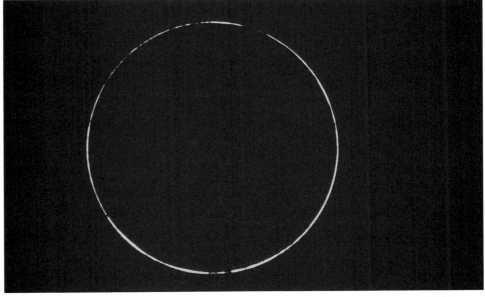

Fig. 14-12. An annular eclipse of the Sun on May 30, 1984. (Courtesy Walter Piorkowski, ©1984.)

✳15✳

The Amateur
Experience

Sharing a dazzling view of Jupiter, the double cluster, or
M31 with a new observer or a child is beyond reward.

Jeffrey L. Hunt,
Cosmic Connections Inc.

So now you have a telescope and some eyepieces. You've had the scope out and you've had a look at a representative sample of what the universe has to offer. Amateur astronomers go through stages in their development. You've gotten interested in astronomy and you've done something about it. You've read this book, a wise choice if I do say so myself, and purchased or made a telescope. You are a sky watcher. As a sky watcher you observe for the sheer pleasure of probing the wonders of the universe. There is no particular pattern to your observing, but that's okay. Remember this is a hobby, something to fill your leisure time. You probably have a full-time job, school, or something that makes more important demands on your time. This is fine, there is nothing wrong with being a sky watcher.

But after a while the simple thrill of sky watching can wear a bit thin. It won't disappear, however, that never really happens. The first clue that something is amiss is when the telescope comes out less often and stays out for less time. When you get around to dragging out the scope, the views still give you that sense of wonder you felt the first time, but now something is missing.

Simply getting the telescope out of the house has turned into a chore, and observing itself seems to have lost its charm and become more like a job. The process of going out each night to find something to look at becomes less attractive because the ''I've seen it all'' syndrome rears its ugly head. As a result, the telescope spends more time in the closet or basement and less time outside under the stars.

I think more telescopes are retired and more amateur astronomers become former amateur astronomers because of simple boredom than anything else. Unfortunately, many amateurs react to this growing boredom by putting an ad in the paper or one of the astronomy magazines and selling the telescope. They solve the problem by eliminating it. If you read the ''Sky-Gazers Exchange'' column in *Sky & Telescope* magazine or the ''Reader Exchange'' section in *Astronomy* magazine, you'll see what I mean. Listing after listing of ''For Sale: Schmidt-Cassegrain 8-inch, only used once (or twice, or whatever),'' followed by a ridiculously low price. Each time I see one of these ads I know that the community of amateur astronomers has lost another of its members to boredom.

It is much better to give your observing some sort of focus, something to concentrate on so that you never have a chance to get bored. Somewhere along the line an observer comes to the realization that in addition to the sheer joy of sky watching he or she may even be able to make a small contribution to science. When that light dawns, amateurs take the first step toward becoming serious observers. Now their observations take a more orderly tack and they are out at their telescopes on a regular basis, though maybe not every night. They make their observations regularly, systematically, and with some concrete goal in mind.

For some it is observing variable stars, for others it is a systematic program directed at a particular planet like Jupiter, or the long task of searching for a comet. The joy of the sky watcher is still there, but it is focused by a desire to learn something more about what is being observed. Some amateurs make the switch from sky watcher to serious observer early in their career, while others take years to make the transition.

Mind you, there is no law that says you must become a serious observer. Some of us never make the change, enjoying our random views of the universe for our entire observing lives. But it is interesting to note that those who become serious observers enjoy what they are doing as much as the beginning sky watcher, if not more. At a public star party you can't tell them apart from the casual sky watcher, and that is where one of the real joys of amateur astronomy lies. Most serious observers have combined their scientific curiosity with the simple joys of sky watching.

THE STARS AND THE SOLITARY OBSERVER

For the most part, amateur astronomy is a solitary avocation. The long hours an amateur spends at the telescope are usually spent alone. Part of the lure of amateur astronomy is, I think, the solitude that it offers the observer. It offers many people an outlet, a chance to escape the pressures of everyday life. But no amateur can consider himself a serious observer and function in a vacuum. That is not how science works, not even amateur science. In order for you to get the most out of amateur astronomy you have to establish lines of communication with other amateurs.

Many amateurs gravitate to groups that encompass their observing specialties. Lunar and planetary observers have the Association of Lunar and Planetary Observers (ALPO). Founded in 1947 by Walter Haas, ALPO is divided into sections devoted to a particular member of the Solar System. Some of the sections are further divided into specialized program areas. For example, the ALPO Lunar Section has programs devoted to: lunar transient phenomena, selected lunar areas, eclipses and photometry, and a lunar dome survey.

Each section or subsection is headed by a recorder who has the responsibility of collecting and analyzing the observations sent in by its members. For a complete list of the ALPO sections and Recorders see Appendix A.

ALPO publishes a journal, *The Journal of the Association of Lunar and Planetary Observers* (JALPO), affectionately known as "The Strolling Astronomer," four times a year. It contains a wealth of information directed at the planetary and lunar observer. ALPO also has a section for the new observer, "The Lunar and Planetary Training Program," that can help a beginner develop his or her observational skills.

Variable star observers have the American Association of Variable Star Observers (AAVSO). Founded in 1911, AAVSO is the clearinghouse for amateur observations of variable stars. AAVSO also is divided into subsections, each with their own area of specialization. Eclipsing binaries, novae and supernovae, Cepheid variables, and the Sun are some of the subsections. AAVSO also puts out a number of publications to help the amateur develop the skills needed to become a working member of the amateur scientific community.

YOUR LOCAL ASTRONOMY CLUB

Probably the best source of assistance for the beginning observer is the local astronomy club. Thousands of amateur astronomers across the United States are members of hundreds of local astronomy clubs. From memberships of over 1,000 to small groups of a dozen or so, the astronomy club is an important part of the growth of the amateur (Fig. 15-1).

Membership in your local club can give you a forum for comparing and discussing observations and techniques. Your membership can also bring additional benefits such as access to equipment that you could never possess on your own. The Southeastern Iowa Astronomy Club (SEIAC), for example, has recently completed a project to refurbish and install a 12-inch Alvan Clark refractor in an observatory built and operated by the club (Fig. 15-2). Work on restoring the 100-year-old instrument took over two years and was done by club members. Today, the club's members, after instruction on the telescope's use and care, have access to an instrument that is a true optical work of art (Fig. 15-3). SEIAC is not unique among astronomy clubs around the country. Many maintain either observatories for the use of their members or observing sites with that most precious of all commodities, a dark sky.

One of the most exciting events held by local groups is the public star party. If you haven't bought a telescope yet, these functions can give you an introduction to the workings of telescopes that you can't get anywhere else. You can compare, side by side, the images produced by reflectors, refractors, SCTs, and a host of weird looking beasts that reach out to the stars and bring them to Earth. You can talk to active amateurs who can help with information on making or buying your telescope. They can also explain the goals of the local group and you can often join the club.

Don't feel that you have to be an astronomical whiz to join your local group. Many members join with little or no knowledge about the field. Amateur astronomy is a growth experience. Members of your local club can help you along the path that best suits your needs and desires.

SHARING THE EXPERIENCE

In order to be truly appreciated, astronomy must be shared. Whether it's through lectures to a senior citizens group, a line of people waiting to see Saturn at the club's star party, or with the kids down the block, astronomy holds no greater thrill for the amateur than when it is shared.

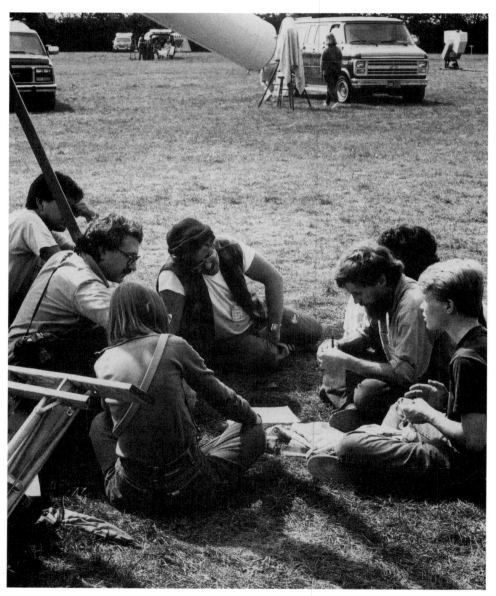

Fig. 15-1. Local astronomy clubs offer you a chance to meet and discuss astronomy with others in your area.

Each month the Chicago Astronomical Society co-hosts a sky watch with Triton College in one of Chicago's suburbs. When I started to attend these I did so out of curiosity and, to be honest, I didn't expect much. The first time I went I was amazed to see over 200 people calmly stand in line at the members' telescopes for a peek at Saturn or the Moon. Because I didn't bring my telescope I had a chance to wander around and listen to some of the comments and discussions going on between the visitors and members.

What I heard was an education in itself. Questions were asked and answered on almost every aspect of telescopes and astronomy that you could imagine. The common ''what power is this thing?'' was asked and answered by a patient member who explained about the importance of image quality over magnification. I heard a couple of kids, they couldn't have been more than 10, ask about the latest theories about lunar crater formation.

With one telescope trained on Saturn, its owner explained that in addition to the rings of the planet the person at the eyepiece could see some of the planet's moons. He explained that the brightest moon, Titan, was visible at about five o'clock in the field. A housewife with a little boy in tow took a look and was puzzled at what she saw. ''How many satellites are visible?'' she asked. The telescope operator promptly replied that with his instrument a person should be able to see five of the moons. She looked back and asked, ''Is that one at about 3 o'clock, close to the rings?'' The operator quickly

Fig. 15-2. The John H. Witte Observatory built and maintained by members of the Southeastern Iowa Astronomy Club (SEIAC). (Courtesy David Bachtell, Southeastern Iowa Astronomy Club, ©1988.)

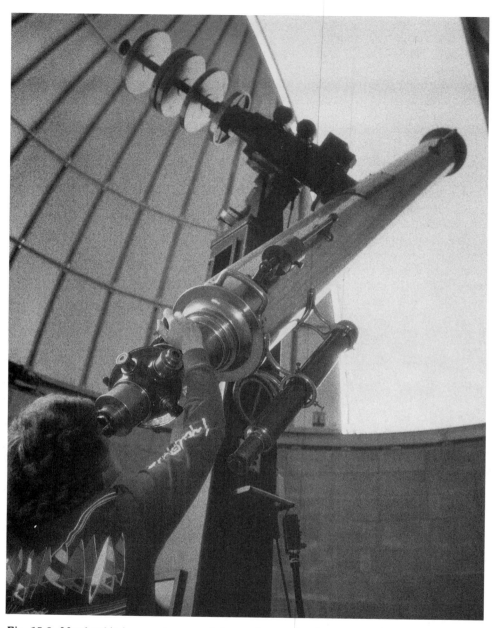

Fig. 15-3. Membership in an astronomy club can give you the chance to use instruments that offer views of the universe unavailable in amateur telescopes. Members of the SEIAC restored this 100-year-old Alvan Clark refractor, an instrument of superb hand-crafted quality. Now members can, after learning the telescope's intricacies, use it for their own projects. (Courtesy David Bachtell, Southeastern Iowa Astronomy Club, ©1988.)

checked and confirmed that Rhea was indeed visible and admitted that he had overlooked it. You don't have to be an amateur astronomer to have an astronomer's eye.

My turn came to man the telescopes when a member wanted a short break. I quickly found myself at a 10-inch reflector. As each person came to the eyepiece I told them that they would be looking at Saturn and to look for as long as they wanted. The line progressed smoothly with each person uttering a comment on the planet's beauty. A little girl moved along with her mother and when it was her turn she slowly mounted the step ladder to the eyepiece. She looked for a moment, said she couldn't see anything, and started down. I told her to take her time and she turned back to the eyepiece. Watching her fidget at the eyepiece I knew that the image of the planet had not hit her eye. Then all of a sudden she froze in place. The look on her face told me she had found Saturn. She looked at her mother and then turned back to the eyepiece where she remained for a long time. All she could say was "Wooow!"

Jeff Hunt is right. That reaction was beyond any reward astronomy had given me prior to that day.

✳A✳

Groups
for the Amateur

AAVSO

The American Association of Variable Star Observers (AAVSO) collects and analyzes information on the brightness changes of variable stars. Information about joining AAVSO may be obtained from:

AAVSO
25 Birch St.
Cambridge, MA 02138
Director: Janet A. Mattei

The AAVSO Solar Division collects information on sunspot activity in the form of sunspot counts. Contact:

AAVSO Solar Division
P. O. Box 8115
Gainesville, FL 32605
Chairman: Peter O. Taylor

The AAVSO Eclipsing Binary Committee is concerned with gaining accurate estimates of light increase and decrease from eclipsing binaries on the night of their predicted eclipse. Estimates are made every 10 minutes or so for a period of 3 or 4 hours on the night of the eclipse. Contact:

AAVSO Eclipsing Binary Division
Route 1
Butlerville, IN 47223
Chairman: Marvin E. Baldwin

AAVSO also has a section devoted to searching for novae and supernovae. The nova searches are carried out using nothing more than an atlas and a pair of 7×50 binoculars. Because the supernova searches are carried out by examining faint galaxies, a telescope of at least medium aperture is required. For more information write:

AAVSO Supernova Division   or   AAVSO Nova Search Division
57 Talbot Rd.
Hazelbrook
New South Wales 2779
Australia
Chairman: Robert O. Evans

AAVSO Nova Search Division
P. O. Box 282
Lewiston, MI 49756
Chairman: Kenneth C. Beckmann

ALPO

The goal of the Association of Lunar and Planetary Observers (ALPO) is to gather information on each member of our Solar System. Each section is concerned with a specific member of the Solar System. Many publish newsletters and information handbooks separate from the Association's journal, *JALPO*. Direct any inquiries about joining ALPO to the Membership Secretary.

Director/Editor
. Lunar Section Recorder
Jupiter Section Asst. Recorder-Eclipse
Timings
John Westfall
P. O. Box 16131
San Francisco, CA 94116

Membership Secretary
Harry D. Jamieson
P. O. Box 143
Heber Springs, AR 72543

Mercury Section Recorder
Richard Baum
25 Whitchurch Rd.
Chester CH3 5QA
England

Venus Section Recorder
Saturn Section Recorder
Lunar Selected Areas Recorder
Julius L. Benton, Jr.
c/o Associates in Astronomy
305 Surrey Rd.
Savannah, GA 31410

Mars Section Recorders
Donald Parker
Jeff Beish
c/o ALPO Mars Recorders
P. O. Box 97-0469
Miami, FL 33197-0469

Jupiter Section Recorders
Phillip W. Budine
P. O. Box 761
Oxford, NY 13830

Jose Olivarez
1469 Valleyview Ct.
Wichita, KS 67212

Remote Planets Recorder
Richard G. Hodgson
Dordt College
Sioux Center, IA 51250

Solar Section Recorders
Richard Hill
4632 E. 14th St.
Tucson, AZ 85711

Solar Section-White Light
Paul Maxson
8839 N. 30th Ave.
Phoenix, AZ 85051

Solar Section-Monochromatic Light
Randy Tatum
3707 Moody
Richmond, VA 23225

Lunar and Planetary Training Program
Jose Olivarez
1469 Valleyview Ct.
Wichita, KS 67212

Solar Section-Eclipses
Lunar Section-Eclipses
Francis Graham
P. O. Box 209
East Pittsburgh, PA 15112

Lunar Transient Phenomena Recorder
Winifred S. Cameron
La Ranchita de la Luna
200 Rojo Dr.
Sedona, AZ 86336

Lunar Dome Survey Recorder
Jim Phillips, M.D.
101 Bull St.
Charleston, SC 29401

Comet Section Recorder
Don E. Machholz
5234 Camden Ave.
San Jose, CA 95124

Meteor Section Recorder
John Griese
963 Elms Common Dr., Apt. 103
Rocky Hill, CT 06067

Minor Planets Section Recorder
Fredrick Pilcher
Illinois College
Jacksonville, IL 62650

THE WEBB SOCIETY

The Webb Society is interested in the objects of the deep sky. These include double stars, clusters, nebulae, and galaxies. Although the Society is based in England, interested observers in the United States may contact:

Ronald J. Morales
1440 S. Marmora Ave.
Tucson, AZ 85713

THE INTERNATIONAL DARK-SKY ASSOCIATION

IDA is a group of astronomers, both amateur and professional, and private individuals who have banned together to do something about the growing problem of light pollution. The aim of the group is public education, promotion of the use of lighting fixtures that preserve the dark skies above the nation's cities, monitoring of sky brightness, and promotion of existing local ordinances that protect the night sky.

Information concerning membership and the resources IDA has available for amateur astronomers and membership can be obtained by contacting:

International Dark-Sky Association
3545 N. Stewart Ave.
Tucson, AZ 85716

B

Upcoming Planetary Oppositions

YEAR	DATE	DEC (DEGREES)	APPARENT DIAMETER (ARCSECONDS)	MAG
MARS				
1990	NOV 27	+22	18.11	−1.8
1993	JAN 7	+26	14.95	−1.2
1995	FEB 12	+18	13.85	−1.0
1997	MAR 17	+04	14.20	−1.0
1999	APR 24	−11	16.18	−1.4
JUPITER				
1989	DEC 27	+23	47.28	−2.3
1991	JAN 29	+19	45.74	−2.1
1992	FEB 29	+09	44.64	−2.0
1993	MAR 30	−02	44.21	−2.0
1994	APR 30	−13	44.53	−2.0
1995	JUN 1	−21	45.54	−2.1
1996	JUL 4	−22	47.05	−2.2
1997	AUG 9	−16	48.64	−2.4
1998	SEP 16	−04	49.70	−2.5
1999	OCT 23	+10	49.70	−2.5
2000	NOV 28	+20	48.63	−2.4

YEAR	DATE	DEC (DEGREES)	APPARENT DIAMETER (ARCSECONDS)	MAG
		SATURN		
1990	JUL 14	−21	18.53	+0.3
1991	JUL 27	−19	18.63	+0.3
1992	AUG 7	−17	18.77	+0.4
1993	AUG 19	−13	18.94	+0.5
1994	SEP 1	−10	19.14	+0.7
1995*	SEP 14	−05	19.37	+0.9
1996*	SEP 26	−00	19.61	+0.7
1997	OCT 10	+04	19.86	+0.4
1998	OCT 23	+08	20.10	+0.2
1999	NOV 6	+13	20.31	+0.0
2000	NOV 19	+17	20.49	−0.1

*Indicates Earth passing through Saturn's ring plane.

The Messier Catalog

Coordinates for Epoch 2000.

M	NGC	R.A. HRS	MIN	DEC DEG	MIN	CONST	MAG	TYPE
1	1952	05	34.5	+22	01	TAU	8.4	DI
2	7089	21	33.5	−00	49	AQR	6.5	GB
3	5272	13	42.2	+28	23	CVn	6.4	GB
4	6121	16	23.6	−26	32	SCO	5.9	GB
5	5904	15	18.6	+02	05	SER	5.8	GB
6	6405	17	40.1	−32	13	SCO	4.2	OC
7	6475	17	53.9	−34	49	SCO	3.3	OC
8	6523	18	03.8	−24	23	SGR	5.8	DI
9	6333	17	19.2	−18	31	OPH	7.9	GB
10	6254	16	57.1	−04	06	OPH	6.6	GB
11	6705	18	51.1	−06	16	SCT	5.8	OC
12	6218	16	47.2	−01	57	OPH	6.6	GB
13	6205	16	41.7	+36	28	HER	5.9	GB
14	6402	17	37.6	−03	15	OPH	7.6	GB
15	7078	21	30.0	+12	10	PEG	6.4	GB
16	6611	18	18.8	−13	47	SER	6.0	OC
17	6618	18	20.8	−16	11	SGR	7.0	DI
18	6613	18	19.9	−17	08	SGR	6.9	OC
19	6273	17	02.6	−26	16	OPH	7.2	GB
20	6514	18	02.6	−23	02	SGR	8.5	DI
21	6531	18	04.6	−22	30	SGR	5.9	OC
22	6656	18	36.4	−23	54	SGR	5.1	GB
23	6494	17	56.8	−19	01	SGR	5.5	OC

M	NGC	R.A.		DEC		CONST	MAG	TYPE
		HRS	MIN	DEG	MIN			
24	———							
25	____	18	31.6	− 19	15	SGR	4.6	OC
26	6694	18	45.2	− 09	24	SCT	8.0	OC
27	6853	19	59.6	+ 22	43	VUL	8.1	PL
28	6626	18	24.5	− 24	52	SGR	6.9	GB
29	6913	20	23.9	+ 38	32	CYG	6.6	OC
30	7099	21	40.4	− 23	11	CAP	7.5	GB
31	224	00	42.7	+ 41	16	AND	3.4	G-S
32	221	00	42.7	+ 40	52	AND	8.2	G-E
33	598	01	33.9	+ 30	39	TRI	5.7	G-S
34	1039	02	42.0	+ 42	47	PER	5.2	OC
35	2168	06	08.9	+ 24	20	GEM	5.1	OC
36	1960	05	36.1	+ 34	08	AUR	6.0	OC
37	2099	05	52.4	+ 32	33	AUR	5.6	OC
38	1912	05	28.7	+ 35	50	AUR	6.4	OC
39	7092	21	32.2	+ 48	26	CYG	4.6	OC
40								
41	2287	06	47.0	− 20	44	CMa	4.5	OC
42	1976	05	35.4	− 05	27	ORI	4.0	DI
43	1982	05	35.6	− 05	16	ORI	9.0	DI
44	2631	08	40.1	+ 19	59	CNC	3.1	OC
45	____	03	47.0	+ 24	07	TAU	1.2	OC
46	2437	07	41.8	− 14	49	PUP	6.1	OC
47	2422	07	36.6	− 14	30	PUP	4.4	OC
48	2548	08	13.8	− 05	48	HYA	5.8	OC
49	4472	12	29.8	+ 08	00	VIR	8.4	G-E
50	2323	07	03.2	− 08	20	MON	5.9	OC
51	5194	13	29.9	+ 47	12	CVn	8.1	G-S
52	7654	23	24.2	+ 61	35	CAS	6.9	OC
53	5024	13	12.9	+ 18	10	COM	7.7	GB
54	6715	18	55.1	− 30	29	SGR	7.7	GB
55	6809	19	40.0	− 30	58	SGR	7.0	GB
56	6779	19	16.6	+ 30	11	LYR	8.2	GB
57	6720	18	53.6	+ 33	02	LYR	9.0	PL
58	4579	12	37.7	+ 11	49	VIR	9.8	G-S
59	4621	12	42.0	+ 11	39	VIR	9.8	G-E
60	4649	12	43.7	+ 11	33	VIR	8.8	G-E
61	4303	12	21.9	+ 04	28	VIR	9.7	G-S
62	6266	17	01.2	− 30	07	OPH	6.6	GB
63	5055	13	15.8	+ 42	02	CVn	8.6	G-S
64	4826	12	56.7	+ 21	41	COM	8.5	G-S
65	3623	11	18.9	+ 13	05	LEO	9.3	G-S
66	3627	11	20.2	+ 12	59	LEO	9.0	G-S

M	NGC	R.A. HRS	MIN	DEC DEG	MIN	CONST	MAG	TYPE
67	2682	08	50.4	+11	49	CNC	6.9	OC
68	4590	12	39.5	−26	45	HYA	8.2	GB
69	6637	18	31.4	−32	21	SGR	7.7	GB
70	6681	18	43.2	−32	18	SGR	8.1	GB
71	6838	19	53.8	+18	47	SGE	8.3	GB
72	6981	20	53.5	−12	32	AQR	9.4	GB
73								
74	628	01	36.7	+15	47	PSC	9.2	G-S
75	6864	20	06.1	−21	55	SGR	8.6	GB
76	6501	01	42.4	+51	34	PER	11.5	PL
77	1068	02	42.7	−00	01	CET	8.8	G-S
78	2068	05	46.7	+00	03	ORI	8.0	DI
79	1904	05	24.5	−24	33	LEP	8.0	GB
80	6093	16	17.0	−22	59	SCO	7.2	GB
81	3031	09	55.6	+69	04	UMa	6.8	G-S
82	3034	09	55.8	+69	41	UMa	8.4	G-Ir
83	5236	13	37.0	−29	52	HYA	7.6	G-S
84	4374	12	25.1	+12	53	VIR	9.3	G-E
85	4382	12	25.4	+18	11	COM	9.2	G-E
86	4406	12	26.2	+12	57	VIR	9.2	G-E
87	4486	12	30.8	+12	24	VIR	8.6	G-E
88	4501	12	32.0	+14	25	COM	9.5	G-S
89	4552	12	35.7	+12	33	VIR	9.8	G-E
90	4569	12	36.8	+13	10	VIR	9.5	G-E
91	4548	12	35.4	+14	30	COM	10.2	G-S
92	6341	17	17.1	+43	08	HER	6.5	GB
93	2447	07	44.6	−23	52	PUP	6.2	OC
94	4736	12	50.9	+41	07	CVn	8.1	G-S
95	3351	10	44.0	+11	42	LEO	9.7	G-S
96	3368	10	46.8	+11	49	LEO	9.2	G-S
97	3587	11	14.8	+55	01	UMa	11.2	PL
98	4192	12	13.8	+14	54	COM	10.1	G-S
99	4254	12	18.8	+14	25	COM	9.8	G-S
100	4321	12	22.9	+15	49	COM	9.4	G-S
101	5457	14	03.2	+54	21	UMa	7.7	G-S
102								
103	581	01	33.2	+60	42	CAS	7.4	OC
104	4594	12	40.0	−11	37	VIR	8.3	G-S
105	3379	10	47.8	+12	35	LEO	9.3	G-E
106	4258	12	19.0	+47	18	CVn	8.3	G-S
107	6171	16	32.5	−13	03	OPH	8.1	GB
108	3556	11	11.5	+55	40	UMa	10.0	G-S
109	3992	11	57.6	+53	23	UMa	9.8	G-S
110	205	00	40.4	+41	41	AND	8.0	G-E

EXPLANATION OF TYPES

Diffuse nebula . . . . .DI
Globular cluster. . . . . .GB
Open cluster . . . . .OC
Planetary nebula . . . . .PL

Galaxy-spiral . . . . .G-S
Galaxy-elliptical . . . . .G-E
Galaxy-irregular . . . . .G-Ir

COMMON NAMES FOR SOME OF THE MESSIER OBJECTS

M1	Crab nebula
M8	Lagoon nebula
M11	The Wild Duck cluster
M13	The Great Hercules cluster
M17	Omega nebula
M20	Triffid nebula
M27	Dumbbell nebula
M31	Andromeda galaxy
M42	Orion nebula

M44	The Beehive cluster or Praesepe
M45	The Pleiades
M51	The Whirlpool galaxy
M57	The Ring nebula
M64	The Blackeye galaxy
M76	The Little Dumbbell nebula
M97	The Owl nebula
M104	The Sombrero galaxy

THE MISSING MESSIERS

I'm sure that you noticed there are a few objects listed with only a number, nothing else. These four objects were mistakes in identification made by Messier when he compiled his original catalog.

M24 is a bright area of the Milky Way in Sagittarius.
M40 is a wide double star in Ursa Major.
M73 is a small asterism in Aquarius.
M102 is a mis-identification of M101.

Deep-Sky Objects For the Telescope

Coordinates for Epoch 2000.

| NGC | R.A. | | DEC | | CONST | MAG | CLASS |
	HRS	MIN	DEG	MIN			
			OPEN CLUSTERS				
457	01	19.1	+58	20	CAS	6.4	I-3-R
663	01	46.0	+61	15	CAS	7.1	III-2-M
752	01	57.8	+37	41	AND	5.7	III-1-M
1857	05	20.2	+39	21	AUR	7.0	II-2-M
2192	06	15.2	+39	51	AUR	10.9	III-1-P
2324	07	04.2	+01	03	MON	8.4	II-2-R
2362	07	18.8	−24	57	CMa	4.1	I-3-P
2432	07	40.9	−19	05	PUP	10.2	II-1-P
2509	08	00.7	−19	04	PUP	9.3	II-1-P
6530	18	04.8	−24	20	SGR	4.6	II-2-M
6611	18	18.8	−13	47	SER	6.0	II-3-M
6819	19	41.3	+40	11	CYG	7.3	II-1-R
7142	21	45.9	+65	48	CEP	9.3	II-2-R
			GLOBULAR CLUSTERS				
5897	15	17.4	−21	01	LIB	8.5	11

NGC	R.A. HRS	MIN	DEC DEG	MIN	CONST	MAG	CLASS

PLANETARY NEBULAE

NGC	HRS	MIN	DEG	MIN	CONST	MAG	CLASS
2392	07	29.2	+20	55	GEM	9.9	irregular disk
3242	10	24.8	−18	38	HYA	8.6	ring structure
6210	16	44.5	+23	49	HER	9.3	smooth disk
6543	17	58.6	+66	38	DRA	8.8	irregular disk
7009	21	04.2	−11	22	AQR	8.3	ring structure
7293	22	29.6	−20	48	AQR	—-	ring structure

BRIGHT NEBULAE

NGC	HRS	MIN	DEG	MIN	CONST	MAG	CLASS
1432	03	46.1	+23	47	TAU	—-	very faint-R
1788	05	06.9	−03	21	ORI	—-	bright-R
1975	05	35.4	−04	41	ORI	—-	bright-E + R
2261	06	39.2	+08	44	MON	—-	bright-E + R
2359	07	18.6	−13	12	CMa	—-	very faint-E
7000	20	58.8	+44	20	CYG	—-	faint-E

GALAXIES

NGC	HRS	MIN	DEG	MIN	CONST	MAG	CLASS
253	00	47.6	−25	17	SCL	7.0	spiral
278	00	52.1	+47	33	CAS	10.8	elliptical
2403	07	36.9	+65	36	CAM	8.9	spiral
4449	12	28.2	+44	06	CVn	9.8	irregular
4494	12	31.4	+25	47	COM	9.8	elliptical
4565	12	36.3	+25	59	COM	10.3	spiral

✳E✳

Double Stars
For the Telescope

What follows is a list of 118 double and multiple stars. Many of them are easily accessible by a small telescope, while others may be beyond an observer's instrument or observing conditions. Each pair is easy to locate. They are all listed by name in the latest edition of *Norton's Star Atlas and Reference Handbook*.

Each star is listed by name with STR being stars from F. G. W. Struve's catalog and OSTR denoting stars from his son Otto's work. Following the name is the constellation in which the star is found, the separation of the pair, and the magnitudes of the components. The position angle (PA) is listed last, with triple stars designated by a T after the PA.

NAME		SEPARATION (arcsec)	MAGNITUDES A	B	POSITION ANGLE
O S T R					
514	AND	5.3	7.0	9.5	168
STR 24	AND	5.2	7.0	8.0	249
STR 79	AND	7.8	6.0	7.0	193
59	AND	16.6	6.5	7.0	35
STR 245	AND	11.2	7.0	8.0	294
STR 2985	AND	15.3	7.0	8.0	254
STR 3042	AND	5.3	7.0	7.0	88
29	AQR	3.9	7.0	7.0	244
41	AQR	5.0	6.0	7.5	114
STR 3008	AQR	4.0	7.0	8.0	175
107	AQR	6.6	5.5	6.5	137

NAME		SEPARATION (arcsec)	MAGNITUDES A	B	POSITION ANGLE
5	AQL	13.0	5.5	7.5	121
STR 2426	AQL	17.1	7.0	8.0	259
STR 2446	AQL	9.6	6.5	8.5	153 (T)
——	—-	34.5	—-	9.5	341
STR 2489	AQL	8.2	6.5	9.5	347
STR 2532	AQL	33.7	6.0	10.0	5
STR 2613	AQL	4.0	7.0	7.5	352
1	ARI	2.7	6.0	7.5	166
Omega	AUR	5.4	5.0	8.0	359
41	AUR	5.5	5.0	7.0	356
Kappa	BOO	13.3	4.5	6.5	236
Pi	BOO	5.6	5.0	6.0	108
Epsilon	BOO	2.9	2.5	5.0	338
39	BOO	2.9	6.0	6.5	45
Xi	BOO	6.9	5.0	7.0	347
STR 419	CAM	3.0	7.0	7.0	75
1	CAM	10.3	5.0	6.0	308
STR 973	CAM	12.5	6.5	7.5	31
STR 1127	CAM	5.3	6.0	8.0	340 (T)
——	—-	11.3	—-	9.0	175
Iota	CNC	30.5	4.5	6.5	307
O S T R					
195	CNC	9.6	7.5	8.0	139
2	CVn	11.4	5.5	8.0	260
STR 1645	CVn	10.2	7.0	7.5	158
Nu-1	CMa	17.5	6.5	8.0	262
STR 3053	CAS	15.2	6.0	7.5	70
Eta	CAS	11.0	3.5	7.5	297
Iota	CAS	2.2	4.0	7.0	241 (T)
——	—-	7.3	—-	8.0	114
Sigma	CAS	3.0	5.5	7.5	326
Kappa	CEP	7.4	4.0	8.0	122
Beta	CEP	13.6	3.5	8.0	250

NAME		SEPARATION (arcsec)	MAGNITUDES A	B	POSITION ANGLE
STR 2819	CEP	12.4	7.5	8.5	57
Xi	CEP	7.6	4.5	6.5	278
Omicron	CEP	3.2	5.5	8.0	214
26	CET	16.0	6.5	9.0	253
66	CET	16.2	6.0	7.5	232
Nu	CET	8.0	5.0	9.5	83
STR 1633	COM	9.0	7.0	7.0	245
24	COM	20.3	5.0	6.5	271
Zeta	CrB	6.3	5.0	6.0	305
O S T R 304	CrB	10.7	6.5	10.0	175
Gamma	CRT	5.2	4.0	9.0	96
Delta	CYG	2.2	3.0	6.5	247
16	CYG	39.0	5.0	5.0	134
17	CYG	26.0	5.0	8.0	70
Psi	CYG	3.1	5.0	7.5	177
52	CYG	6.6	4.0	9.0	67
STR 2725	DEL	5.7	7.5	8.0	9
Gamma	DEL	10.1	4.5	5.0	268
Eta	DRA	5.3	3.0	8.0	143
17	DRA	3.2	5.5	6.5	109
Epsilon	EQU	10.6	5.5	7.0	70
20	GEM	20.0	6.0	7.0	210
38	GEM	7.0	5.5	7.5	151
Lambda	GEM	9.6	3.0	10.0	33
Delta	GEM	6.3	3.5	8.0	218
Kappa	GEM	7.0	4.0	10.0	239
Kappa	HER	28.2	5.0	6.0	12
Alpha	HER	4.6	3.0	5.5	110
Delta	HER	8.8	3.0	8.5	241
Rho	HER	4.1	4.5	5.5	317
95	HER	6.3	5.0	5.0	258
100	HER	14.2	6.0	6.0	183

NAME		SEPARATION (arcsec)	MAGNITUDES		POSITION ANGLE
			A	B	
17	HYA	4.3	7.0	7.0	359
Theta	HYA	29.4	4.0	10.0	197
N	HYA	9.1	6.0	6.0	210
54	HYA	8.8	5.0	7.0	126
STR 1360	LEO	14.2	7.5	8.0	242
54	LEO	6.5	4.5	6.5	110
88	LEO	15.4	6.5	8.5	326
18	LIB	19.6	6.0	9.5	39
STR 1962	LIB	11.8	6.5	6.5	189
19	LYN	14.7	5.5	6.5	315
20	LYN	15.1	7.0	7.0	254
Zeta	LYR	43.7	4.5	5.5	150
STR 2470	LYR	13.5	6.5	8.0	271
Eta	LYR	28.3	4.0	8.0	83
8	MON	13.2	4.5	6.5	27
Beta	MON	7.4	5.0	5.5	132
STR 926	MON	11.7	7.5	8.5	289
Zeta	MON	32.0	5.0	10.5	245
19	OPH	23.2	6.0	9.5	89
39	OPH	10.8	6.0	7.0	355
61	OPH	20.6	6.0	6.5	93
STR 627	ORI	21.0	6.5	7.0	260
Rho	ORI	7.0	4.5	8.5	63
23	ORI	32.0	5.0	7.0	28
Lambda	ORI	4.4	4.0	6.0	44
Iota	ORI	11.4	3.0	7.0	141
Theta	PER	18.3	4.0	10.0	303
Eta	PER	28.4	4.0	8.5	301
40	PER	20.0	5.0	9.5	238
Zeta	PER	12.9	3.0	9.5	209
Epsilon	PER	8.8	3.0	8.0	9
35	PSC	11.8	6.0	7.5	149
Zeta	PSC	23.6	4.5	5.5	63

	NAME	SEPARATION (arcsec)	MAGNITUDES A	B	POSITION ANGLE
Theta	SGE	11.9	6.5	8.5	325
Sigma	SCO	20.0	3.0	9.0	273
Delta	SER	3.9	4.0	5.0	178
6	SER	3.1	5.5	9.5	20
Theta	SER	22.2	4.5	5.0	103
STR 430	TAU	26.1	6.0	9.0	55 (T)
——	—-	37.0	—-	10.0	301
Chi	TAU	19.5	5.5	7.5	25
Iota	TRI	3.8	5.0	6.5	72
STR 232	TRI	6.5	7.5	7.5	246
Nu	UMa	7.2	4.0	10.0	147
Gamma	VIR	4.7	3.5	3.5	306
STR 2769	VUL	17.9	6.5	7.5	300

Variable Stars
For the Telescope

DESIGN	NAME	RANGE	PERIOD	TYPE
*010884	RU CEP	8.2 – 9.4	109d	SR
015254	U PER	8.1 – 11.3	321d	LPV
*0214 – 03	Omicron CETI	3.6 – 9.1	332d	LPV
021558	S PER	8.6 – 10.6	?	SR
*0243 – 12	Z ERI	7.0 – 8.6	80d	SR
*0247 – 08	RR ERI	7.4 – 8.6	97d	SR
032043	Y PER	8.4 – 10.3	253d	LPV
0455 – 14	R LEP	6.8 – 9.8	421d	LPV
*0506 – 11	RX LEP	5.0 – 7.0	?	IRR
*052504	CK ORI	5.9 – 7.1	120d?	SR
053068	S CAM	7.7 – 11.1	327d	SR
053613	FX ORI	8.2 – 10.4	720d	SR
*054907	Alpha ORI	0.4 – 1.3	2070d	SR
072046	Y LYN	7.8 – 10.3	110d	SR
081112	R CNC	6.8 – 11.2	361d	LPV
*0824 – 05	RT HYA	7.5 – 9.2	253d	SR
085120	T CNC	7.6 – 10.5	482d	SR
*085211	RT CNC	8.3 – 9.4	90d	SR
094211	R LEO	5.8 – 10.0	312d	LPV
*112245	ST UMa	7.7 – 9.5	81d?	SR
115158	Z UMa	7.9 – 10.8	196d	SR

*Indicates binocular variable.

DESIGN	NAME	RANGE	PERIOD	TYPE
*122001	SS VIR	6.8 – 8.9	355d	LPV
123307	R VIR	6.9 – 11.5	146d	LPV
123961	S UMa	7.8 – 11.7	226d	1pv
*131546	V CVn	6.7 – 8.8	192d	SR
*1324 – 22	R HYA	4.5 – 9.5	390d	LPV
142539	V BOO	7.0 – 12.0	258d	SR
153738	RR CrB	8.4 – 10.1	60d	SR
*155947	X HER	7.5 – 8.6	95d	SR
1621 – 12	V OPH	7.5 – 10.2	298d	LPV
*162542	g HER	5.7 – 7.2	70d	SR
163172	R UMi	9.1 – 10.4	324d	SR
164715	S HER	7.6 – 12.6	307d	LPV
181136	W LYR	7.9 – 12.2	197d	LPV
*183308	X OPH	6.8 – 8.8	334d	LPV
*185243	R LYR	3.9 – 5.0	46d	LPV
190108	R AQL	6.1 – 11.5	291d	LPV
194048	RT CYG	7.3 – 11.8	190d	LPV
*200938	RS CYG	7.2 – 9.0	417d	SR
201647	U CYG	7.2 – 10.7	462d	LPV
203816	S DEL	8.8 – 12.0	277d	LPV
*204017	U DEL	7.6 – 8.9	?	SR
210868	T CEP	6.0 – 10.3	388d	LPV
*214058	Mu CEP	3.6 – 5.1	?	SR
231040	TY AND	8.8 – 10.5	260d	SR
233335	ST AND	8.8 – 11.1	328d	SR
2338 – 15	R AQR	6.5 – 10.3	387d	LPV

*Indicates binocular variable.

Constellations
Visible From 40° N

NAME	ABBREVIATION	SEASON VISIBLE
Andromeda	And	Fall
Antila	Ant	Spring
*Aquarius	Aqr	Spring
Aquila	Aql	Summer
*Aries	Ari	Fall
Auriga	Aur	Winter
Bootes	Boo	Summer
Caelum	Cae	Winter
Camelopardalis	Cam	Winter
*Cancer	Cnc	Spring
Canis Venatici	CVn	Spring
Canis Major	CMa	Winter
Canis Minor	CMi	Winter
*Capricornus	Cap	Fall
Cassiopeia	Cas	Fall
Centaurus	Cen	Summer
Cephus	Cep	Fall
Cetus	Cet	Fall

*Indicates zodiacal constellation.

NAME	ABBREVIATION	SEASON VISIBLE
Columba	Col	Winter
Coma Bernices	Com	Spring
Corona Austrais	CrA	Summer
Corona Borealis	CrB	Summer
Corvus	Crv	Spring
Crater	Crt	Spring
Cygnus	Cyg	Summer
Delphinus	Del	Summer
Draco	Dra	Summer
Equuleus	Equ	Fall
Eridanus	Eri	Winter
Fornax	For	Fall
*Gemini	Gem	Winter
Grus	Gru	Fall
Hercules	Her	Summer
Hydra	Hya	Spring
Lacerta	Lac	Fall
*Leo	Leo	Spring
Leo Minor	LMi	Spring
Lepus	Lep	Winter
*Libra	Lib	Summer
Lupus	Lup	Summer
Lynx	Lyn	Winter
Lyra	Lyr	Summer
Microscopium	Mic	Summer
Monoceros	Mon	Winter
Ophiuchus	Oph	Summer
Orion	Ori	Winter
Pegasus	Peg	Fall
Perseus	Per	Fall
*Pisces	Psc	Fall
Pisces Austrinus	PsA	Fall

*Indicates zodiacal constellation.

NAME	ABBREVIATION	SEASON VISIBLE
Puppis	Pup	Winter
Pyxis	Pyx	Spring
Sagitta	Sge	Summer
*Sagittarius	Sgr	Summer
*Scorpius	Sco	Summer
Sculptor	Scl	Fall
Scutum	Sct	Summer
Serpens	Ser	Summer
Sextans	Sex	Spring
*Taurus	Tau	Winter
Triangulun	Tri	Fall
Ursa Major	UMa	Spring
Ursa Minor	UMi	Summer
*Virgo	Vir	Spring
Vulpecula	Vul	Summer

*Indicates zodiacal constellation.

The Greek Alphabet

Alpha	α	Iota	ι	Rho	ϱ
Beta	β	Kappa	$\varkappa$	Sigma	σ
Gamma	γ	Lambda	λ	Tau	τ
Delta	δ	Mu	μ	Upsilon	υ
Epsilon	ϵ	Nu	ν	Phi	ϕ
Zeta	ζ	Xi	ξ	Chi	χ
Eta	η	Omicron	o	Psi	ψ
Theta	θ	Pi	π	Omega	ω

✴I✴

Astronomical Suppliers

Finding good astronomical equipment can be more difficult than finding a magnitude 15 galaxy from the center of Chicago. When you look for an equipment manufacturer or dealer, don't look only at price. These people can be fountains of information for the new observer. Once you find one, treat him with the respect of a fine eyepiece.

TELESCOPE MANUFACTURERS

Astro Physics
7470 Forest Hills Rd.
Loves Park, IL 61111

Manufactures a line of refractors, mounts, and accessories. Free catalog.

Celestron International
Box 3578
Torrance, CA 90503

Manufactures a complete line of refractor, reflector, and catadioptric telescopes and accessories.

Coulter Optics
P. O. Box K
Idyllwild, CA 92349

Manufactures a line of dobsonian telescopes, optics, and accessories. Free catalog.

Meade Instruments
1675 Toronto Way
Costa Mesa, CA 92626

Manufactures a complete line of refractor, reflector, and catadioptric telescopes, optics, and accessories.

Questar Corp.
P. O. Box 59
New Hope, PA 18938

Manufactures a line of maksutov telescopes.

Star-Liner
1106 S. Columbus Blvd.
Tucson, AZ 85711

Manufactures a line of reflector telescopes and optics.

Tele-Vue Optics
20 Dexter Plaza
Pearl River, NY 10965

Manufactures a line of refractors, eyepieces, mountings, and accessories.

Vernonscope & Co.
Candor, NY 13743

Manufactures a line of refractors, eyepieces, filters, and accessories. Free catalog.

DEALERS: RETAIL AND MAIL ORDER

Many of these dealers distribute equipment manufactured by the companies listed above. In addition, many distribute their own brand of telescopes and accessories.

Ad-Libs Astronomics
2401 Tee Cir.
Suite 105
Norman, OK 73069

Astro-Tech
101 W. Main
Box 2001
Ardmore, OK 73402

Cosmic Connections
1460 N. Farnsworth Ave.
Aurora, IL 60505

Chicago Optical and Supply
Box 1361
Morton Grove, IL 60053

Edmund Scientific
101 E. Glouchester Pike
Barrington, NJ 08012

Northern Sky Telescopes
5667 Duluth St.
Golden Valley, MN 55422

Parks Company
270 Easy St.
Simi Valet, CA 93065

Orion Telescope Center
P. O. Box 1158
Santa Cruz, CA 95061

Pauli Discount Optics
29 Kingswood Rd.
Danbury, CT 06811

Roger W. Tuthill
Box 1086
Mountainside, NJ 07092

University Optics
P. O. Box 1205
Ann Arbor, MI 48106

ACCESSORIES, FILTERS, AND TELESCOPE-MAKING SUPPLIES

Many amateurs prefer to build their own telescopes from scratch or from commercially produced parts. These companies supply such things as mirrors, telescope components, and accessories.

DayStar Filter Company
P. O. Box 1290
Pomona, CA 91769

Specializes in solar filters and nebular filters.

E & W Optical
2420 E. Hennepin Ave.
Minneapolis, MN 55413

Specializes in precision optics and repairs.

A. Jaegers
691 Merrick Rd.
Lynbrook, NY 11563

Specializes in precision optical components, telescope making supplies, and accessories.

Lorraine Precision Optics
724 Florence Ave., #2
Evanston, IL 60202

Specializes in precision optical systems for tri-schiefspiegler telescopes.

Lumicon
2111 Research Dr., #5
Livermore, CA 94550

Specializes in accessories, light pollution and nebular filters, and astrophoto accessories.

Kenneth Novak & Co.
Box 69
Ladysmith, WI 54848

Specializes in telescope making accessories.

Sky Optical
12428 Gladstone
Sylmar, CA 91342

Specializes in precision optics, telescopes, and telescope making supplies.

Telescopics
P. O. Box 98
La Canada, CA 91011

Specializes in precision optics and telescopes.

Thousand Oaks Optical
Box 248098
Farmington, MI 48332

Specializes in precision full-aperture solar filters.

PUBLISHERS AND BOOK DEALERS

Abrams Planetarium
Michigan State University
East Lansing, MI 48824

Publishes monthly *Sky Calendar*.

AstroMedia
1027 N. Seventh St.
Milwaukee, WI 53233

Publishes astronomical books and the following publications: *Astronomy Magazine* (monthly), *Deep Sky Magazine* (quarterly), *Telescope Making* (quarterly), and *Odyssey* (monthly).

Herbert A. Luft
46 Woodcrest Dr.
Scotia, NY 12302

Distributes astronomical books.

Sky Publishing Corp.
Box 9111
Belmont, MA 02178-9918

Publishers of astronomical books and *Sky & Telescope* magazine (monthly).

Willmann-Bell
P. O. Box 35025
Richmond, VA 23235

Publishers of astronomical books.

Bibliography

GENERAL ASTRONOMY

Abell, George O. *Exploration of the Universe*. New York: Holt, Reinhart & Winston, 1974.

_____. *Realm of the Universe*. Philadephia: Saunders College Publishing, 1984.

Burgess, Eric. *Celestial Basic*. Berkeley, Calif.: Sybex, 1982.

Corliss, W. R. *Mysterious Universe: A Handbook of Astronomical Anomalies*. Glen Arm, Md.: The Sourcebook Project, 1979.

Jastrow, R. & M. H. Thompson. *Astronomy: Fundamentals & Frontiers*. New York: John Wiley & Sons, 1984.

ASTRONOMICAL HISTORY

Allen, R. H. *Star Names: Their Lore and Meaning*. New York: Dover Publications, 1963.

Ashbrook, Joseph. *The Astronomical Scrapbook*. Cambridge, Mass.: Sky Publishing Corp., 1984.

Clerke, Agnes M. *A Popular History of Astronomy During the Nineteenth Century*. London: Adam & Charles Black, 1893.

_____. *The Herschels and Modern Astronomy*. New York: MacMillan, 1895.

Dreyer, J. L. E. *A History of Astronomy from Thales to Kepler*. New York: Dover Publications, 1953.

Hoskin, M. A. *William Herschel and the Construction of the Universe*. London: Oldbourne, 1963.

Lowell, A. Lawrence. *Biography of Percival Lowell*. New York: MacMillan, 1935.

MacPherson, Hector. *Herschel*. New York: MacMillan, 1919.

Thiel, Rudolf. *And There Was Light*. New York: Alfred A. Knopf, 1957.

Smyth, Adm. William H. *The Bedford Catalogue*. Richmond, Va.: Willmann-Bell, 1981.

AMATEUR GUIDES

Fulton, Ken. *The Light-Hearted Astronomer*. Milwaukee, Wis.: Astromedia, 1984.

MacRobert, Alan. *Backyard Astronomy*. Cambridge, Mass.: Sky Publishing Corp., 1986.

Moore, Patrick. *The Amateur Astronomer*. London: Lutterworth Press, 1974.

_____. *Practical Amateur Astronomy*. London: Lutterworth Press, 1973.

Peltier, Leslie. *Starlight Nights*. Cambridge, Mass.: Sky Publishing Corp., 1965.

Roth, Gunter. *The Amateur and His Telescope*. New York: Van Norstrand, 1963.

Sherrod, P. Clay. *A Complete Manual of Amateur Astronomy*. New York: Prentice-Hall, 1981.

Sidgwick, J. B. *Introducing Astronomy*. London: Faber & Faber, 1951.

_____. *Observational Astronomy for Amateurs*. New York: Dover Publications, 1980.

LEARNING THE SKY

Beyer, Steven L. *The Star Guide*. Boston: Little Brown, 1986.

Burnham, Robert. *The Star Book. Milwaukee, Wis.: Astromedia, 1983.*

Cleminshaw, C. H. *The Beginner's Guide to the Skies*. New York: Thomas Crowell, 1977.

Gallant, Roy A. *The Constellations: How They Came to Be*. New York: Four Winds Press, 1979.

Kyselka, W. & R. Lanterman. *North Star to Southern Cross*. Honolulu: Univ. of Hawaii Press, 1976.

Mayall, R. & M. Mayall. *Olcott's Field Book of the Skies*. New York: G.P. Putnam's Sons, 1954.

Nealy, Henry M. *The Stars by Clock and Fist*. New York: Viking, 1972.

Peltier, Leslie. *Guideposts to the Stars: Exploring the Skies throughout the Year*. New York: MacMillan, 1972.

Phillips, Henry. "Learning the Constellations: Part 1: Winter." *Astronomy* (December 1977): 42.

_____. "Part 2: Spring." *Astronomy*, (April 1978): 36

_____. "Part 3: Summer." *Astronomy*, (July 1978): 36

_____. "Part 4: Autumn." *Astronomy*, (October 1978): 46.

Raymo, Chet. *365 Starry Nights*. Englewood Cliffs, N.J.: Prentice-Hall, 1982.

Whitney, Charles. *Whitney's Star Finder*. New York: Alfred A. Knopf, 1985.

SEEING, TELESCOPES, AND OPTICS

Asimov, Isaac. *Eyes on the Universe: A History of the Telescope*. Boston: Houghton Mifflin, 1975.

Bell, Louis. *The Telescope*. New York: Dover Publications, 1981.

Bowen, Keith. "Vision and the Amateur Astronomer. *Sky & Telescope* (April 1984): 321

Burke, Paul. "Alt-azimuth Setting Circles." *Astronomy*, (April 1982): 71.

Burton, Bill. "Ready, Compute, Aim!" *Astronomy*, (July 1988): 71.

Gordon, Rodger. "Resolution and Contrast." *JALPO*, (March 1979): 180.

Howard, Neale. *Standard Handbook for Telescope Making*. New York: Harper & Row, 1984.

Ingalls, Albert, ed. *Amateur Telescope Making*. New York: Scientific American, 3 vols., 1978.

King, Henry C. *The History of the Telescope*. New York: Dover Publications, 1955.

Masland, Richard H. "The Functional Architecture of the Retina." *Scientific American*, (Vol. 255 No. 6): 102.

Minnaert, M. *The Nature of Light and Colors in the Open Air*. New York: Dover Publications, 1954.

Sheehan, William. "On an Observation of Saturn: The eye and the Astronomical Observer." *JALPO*, (August 1980): 150.

Shumaker, Bryan. "Build a Binocular 'Sky Scanner'." *Astronomy*, (February 1988): 64.

Sidgwick, J. B. *Amateur Astronomer's Handbook*. New York: Dover Publications, 1980.

Texereau, Jean. *How to Make a Telescope*. Richmond, V.A.: Willmann-Bell, 1984.

Treisman, Anne. "Features and Objects in Visual Processing." *Scientific American*, (Vol. 255 No. 5): 114.

OBSERVING THE MOON: GENERAL

Baldwin, Ralph. *The Face of the Moon*. Chicago: Univ. of Chicago Press, 1947.

Benton, Julius L., Jr. *The ALPO Lunar Selected Areas Program: A Manual for Visual Observations*. ALPO, 1983.

British Astronomical Association. *Guide to Observing the Moon*. Hillside, N.J.: Enslow Publishing, 1986.

Cameron, Winifred S. "Results from the LTP program for the ALPO." *JALPO*, (March 1979): 195.

_____."Results from the LTP Program for the ALPO—concluded." *JALPO*, (June 1979): 328.

_____. "Report on the ALPO Lunar Transient Phenomena Program." *JALPO*, (September 1974): 1.

Cherrington, E. H. *Exploring the Moon through Binoculars and a Small Telescope*. New York: Dover Publications, 1984.

Corliss, W. R. *The Moon and Planets: A Catalog of Astronomical Anomalies*. Glen Arm, M.D.: The Sourcebook Project, 1979.

Dunham, J. B. *The IOTA Observer's Manual*. IOTA, 1985.

Huddleston, Marvin. "The ALPO Lunar Selected Areas Program." *JALPO*, (November 1976): 113.

Kopal, Zdenek. *The Moon in the Post-Apollo Era*. Boston: Reidel Publishing, 1974.

Kozyrev, Nikolai. "Observation of a Volcanic Process on the Moon." *Sky & Telescope*, (February 1959): 184.

Moore, Patrick. *New Guide to the Moon*. New York: W. W. Norton, 1976.

Povemire, Harold. *Graze Observer's Handbook*. New York: Vantage, 1975.

Schultz, Peter H. *Moon Morphology: Interpretations Based on Lunar Orbiter Photography*. Austin, Tex.: Univ. of Texas Press, 1976.

Wilkins, H. P. & P. Moore. *The Moon*. New York: MacMillan, 1960.

Westfall, John. "A Comprehensive Catalog of Lunar Transient Phenomena." *JALPO*, (March 1979): 192.

_____. "A Generic Classification of Lunar Domes." *JALPO*, (January 1964).

OBSERVING THE MOON: ATLASES

Alter, Dinsmore. *Pictorial Guide to the Moon*. New York: Thomas Crowell, 1963.

Bowker, D. E. & J. K. Hughes. *Lunar Orbiter Photographic Atlas of the Moon*. Washington, D.C.: NASA (SP-206), 1971.

Kopal, Zdenek. *A New Photographic Atlas of the Moon*. New York: Taplinger Publishing, 1971.

Kosofsky, L. J. & F. El-Baz. *The Moon as Viewed by Lunar Orbiter*. Washington, D.C.: NASA (SP-200), 1970.

Kriegers, John Nep. *Mond Atlas: Neue Folge*. London: Wm. Wesley & Son, 1912.

Kuiper, G. P., et al. *Consolidated Lunar Atlas*. Univ. of Arizona Lunar & Planetary Lab, 1967.

_____. *Rectified Lunar Atlas*. United States Air Force.

_____. *Atlas & Gazetteer of the Nearside of the Moon*. Washington, D.C.: NASA (SP-241), 1971.

OSERVING THE PLANETS: GENERAL

Briggs, G. A. & F. W. Taylor. *The Cambridge Photographic Atlas of the Planets*. Cambridge, Mass.: Cambridge Univ. Press, 1982.

Dobbins, T., D. Parker, & C. Capen. *Introduction to Observing and Photographing the Solar System*. Richmond, V.A.: Willmann-Bell, 1988.

Doherty, Paul. *Atlas of the Planets*. New York: McGraw Hill, 1980.

Eastman Kodak Co. *Kodak Filters for Scientific and Technical Use*. Rochester, N.Y.: Eastman Kodak, 1970.

Moore, Patrick. *The New Guide to the Planets*. New York: W. W. Norton, 1971.

Roth, Gunter. *Handbook for Planet Observers*. London: Faber & Faber, 1970.

Webb, Rev. T. W. *Celestial Objects for Common Telescopes (Vol. 1)*. New York: Dover Publications, 1963.

Budine, Phillip. "Observer's Guide to the Planet Jupiter." *JALPO*, (October 1985): 56.

_____. "Procedures for Recording CM Transits on Jupiter and Their Reduction." *JALPO*, (March 1970): 37.

Mackal, P. Karl. "Synoptic Report for the 1980-81 Apparition of Jupiter." *JALPO*, (February 1985): 2.

_____. Concluded. *JALPO*, (October 1985): 64.

Peek, Bertand. *The Planet Jupiter*. London: Faber & Faber, 1958.

MARS

Batson, R. M., P. M. Bridges, & J. L. Inge. *Atlas of Mars*. Washington, D.C.: NASA, 1979.

Beish, J. "The 1988 Perihelic Apparition of Mars." *JALPO*, (March 1988): 158.

Beish, J & C. Capen. *The Mars Observer's Handbook*. Pasadena, Calif.: The Planetary Society.

Capen, C., D. Parker, & J. Beish. "A Martian Observer's Menu for 1986." *JALPO*, (July 1986): 183.

Capen, Virginia. "A history of Martian Nomenclature." *Astronomy*, (April 1975): 20.

Glasstone, S. *The Book of Mars*. Washington, D.C.: NASA (SP-179), 1968.

Lowell, Percival. *Mars*. Boston: Houghton Mifflin, 1895.

————————. *Mars and its Canals*. New York: MacMillan, 1908.

Mutch, T., et al. *The Geology of Mars*. Princeton, N.J.: Princeton Univ. Press, 1976.

Pickering, William H. *Mars*. Boston: The Gorham Press, 1921.

Slipher, E. C. *A Photographic History of Mars: 1905-1961*. Lowell Observatory, 1962.

————————. *Viking Orbiter Views of Mars*. Washington, D.C.: NASA (SP-441), 1980.

————————. *Mars as Viewed by Mariner 9*. Washington, D.C.: NASA (SP-329), 1974.

SATURN

Alexander, A. F. *The Planet Saturn*. New York: Dover Publications, 1962.

Benton, Julius L. Jr., *The Saturn Handbook*. ALPO, 1975.

————————. "Southern Globe and Ring Features of Saturn: A Summary Analysis (1966-1980)." *JALPO*, (July 1986): 190.

————————. "The 1985-86 Apparition of Saturn." *JALPO*, (August 1988): 197.

VENUS

Benton, Julius L., Jr. *The Venus Handbook*. ALPO, 1988.

————————. "Two Eastern (Evening) Apparitions of the Planet Venus." *JALPO* (November 1976): 100.

————————. *"Three Western (Morning) Apparitions of the Planet Venus."* *JALPO*, (October 1985): 78.

OBSERVING COMETS AND METEORS

Edberg, Stephan. *International Halley Watch Amateur Observers Manual for Scientific Comet Studies*. Hillside, N.J.: Enslow Publishing, 1983.

Edberg, S. & D. Levy. *Observe Comets*. Washington, D.C.: Astronomical League, 1985.

Kronk, G. W. *Meteor Showers: A Descriptive Catalog*. Hillside, N.J.: Enslow Publishers, 1988.

Peltier, Leslie. "Comet Hunting for Amateurs." *Sky & Telescope*, (January 1962): 10.

Whipple, Fred. *The Mystery of Comets*. Washington, D.C.: Smithsonian Institute Press, 1985.

Observing Double Stars

Burnham, S. W. *A General Catalog of Double Stars: Part II—Notes to the Catalog*. Washington, D.C., The Carnegie Institution of Washingtion, 1906.

Couteau, Paul. *Observing Visual Double Stars*. Cambridge, Mass.: MIT Press, 1981.

Crossley, E., J. Gledhill, & J. Wilson. *Handbook of Double Stars*. London: MacMillan & Co., 1879.

Pither, Colin. "Measuring Double Stars with a Grating Micrometer." *Sky & Telescope*, (June 1980): 519.

Peterson, Harold. "Double Stars in a 3-inch Refractor." *Sky & Telescope*, (September 1954): 396.

Observing Variable Stars

Ford, Clinton. "Sidelights on Variable Star Observing." *Sky & Telescope*, (December 1961): 312.

Furness, Caroline. *An Introduction to the Study of Variable Stars*. Boston: Houghton Mifflin, 1915.

Mayall, Margaret. *Manual for Observing Variable Stars*. Cambridge, Mass.: AAVSO, 1970.

Merrill, Paul. *The Nature of Variable Stars*. New York: MacMillan, 1938.

Observing Deep-Sky Objects

_____. Berry, Richard. "Nebular Filters." *Astronomy*, (March 1979): 51.

_____. "More about Nebular Filters." *Astronomy*, (August 1979): 46.

di Cicco, Dennis. "Filters That Pierce the Nighttime Veil." *Sky & Telescope*, (March 1979): 231.

Eicher, David. "21 Challenging Deep Sky Objects." *Astronomy*, (September 1980): 38.

Hartung, E. J. *Astronomical Objects for Southern Telescopes*. Cambridge, Mass.: Cambridge Univ. Press, 1984.

Jones, Kenneth Glyn, ed. *Webb Society Deep-Sky Observer's Handbook* (7 vols.). Hillside, N.J.: Enslow Publishers, 1986.

_____. *Messier's Nebula and Star Clusters*. American Elsevier Publishing Co, 1968.

_____. *The Search for the Nebulae*. London: Alpha Academic, 1975.

Kratz, Dave. "Observing Planetary Nebulae." *Deep Sky Magazine*, (Spring 1984): 6.

Ling, Alister. "Observing Planetaries from the City." *Deep Sky Magazine*, (Summer 1986): 26.

Mallas, John & E. Kreimer. *The Messier Album*. Cambridge, Mass.: Sky Publishing Corp., 1978.

Morales, Ronald. *The Amateur Astronomer's Catalog of 500 Deep Sky Objects (Vol. 1)*. Tucson, Ariz.: Aztex Corp., 1986.

Schur, Chris. "Observing Small Nebulae Part 1." *Deep Sky Magazine*, (Winter 1984): 24.

_____. ."Part 2." *Deep Sky Magazine*, (Summer 1985): 20.

Thompson, G. & J. Bryan. "Making Visual Discoveries of Supernovae." *Deep Sky Magazine*, (Winter 1983): 16.

Webb, Rev. T. W. *Celestial Objects for Common Telescopes* (Vol. 2). New York: Dover Publications, 1963.

OBSERVING THE SUN

Baxter, W. M. *The Sun and the Amateur Astronomer*. London: David & Charles, 1963.

Bray, R. J. & R. E. Loughhead. *Sunspots*. New York: Dover Publications, 1982.

Hill, Richard. *Handbook for the White Light Observation of Solar Phenomena*. ALPO, 1984.

Maag, R., J. Sherlin, & R. van Zandt. *Observe and Understand the Sun*. Washington, D.C.: Astronomical League, 1976.

Nicolson, Ian. *The Sun*. Chicago: Rand McNally, 1982.

Noyes, Robert W. *The Sun, Our Star*. Cambridge, Mass.: Harvard Univ. Press, 1982.

STAR ATLASES AND CATALOGS

Becvar, A. *Skalnate Pleso Atlas of the Heavens*. Cambridge, Mass.: Sky Publishing Corp., 1958.

_____. *Atlas of the Heavens Catalogue*. Cambridge, Mass.: Sky Publishing Corp., 1958.

Burnham, R. J., Jr. *Burnham's Celestial Handbook*. New York: Dover Publications, 1978.

Hirshfeld, A. & R. Sinnot. *Sky Catalogue 2000* (2 vols.). Cambridge, Mass.: Sky Publishing Corp., 1985.

Norton, A. P. *Star Atlas and Reference Handbook, 17th Edition*. Essex, England: Longman Scientific & Technical, 1978.

Sarna & Assoc. "Deep Sky Detail Charts." *Sarna Deep Sky Atlas*. Downers Grove, Ill., 1981.

Smithsonian Astrophysical Observatory. *Star Atlas of Reference Stars and Nonstellar Objects Star Catalog* (4 vols.). Washington, D.C.: Smithsonian Institute, 1971.

Tirion, W. *Sky Atlas 2000*. Cambridge, Mass.: Sky Publishing Corp., 1981.

Tirion, W., B. Rappaport, & G. Lovi. *Uranometria 2000* (2 vols.). Richmond, V.A.: Willmann-Bell, 1987.

Index

reflection nebulae, 37, 262, 264
refracting telescopes, 2, 39, 40, 45-50, 146, 147
relative spot number (RSN), 283
resolution, 15, 42, 68, 213, 214, 217
limits to, 44
resolving power, 42, 68, 213, 214, 217
reversed orientation, 95
Revised NGC . . . Nonstellar Objects, 253
Rhea, 190
Rheita crater, 123, 124, 131
Riccioli, Giovanni, 155, 209
rich-field telescope, 52, 53, 70, 248
richness, clusters, 256
ridges, lunar, 118
right ascension, 81, 83, 86, 90
right declination, 81
rilles, 118
Rima Posidonius II, 126
Rima Tenuis, 7, 8, 163
Ring Nebula, 99, 264-266, 271
ring plain craters, 117
rings
Cecropia, 163
Ierne, 163
Lemuria, 163
planetary, 29, 42
Saturnian, 188-190
roof prism, binoculars, 23, 25
RR Lyrae stars, 229, 230
RV Tauri stars, 229

S

Sagittarius, 262
salenography, 112
Sarna Deep-Sky Atlas, 249, 252
satellites, 28
Jovian, 173, 184-188
Saturnian, 189-190
Saturn, 5, 29, 103, 141, 142, 145, 188-190, 304
rings of, 188-190
satellites of, 189-190
scarps, 124
lunar, 118
Schiaparelli, Giovanni, 7, 159, 166, 167
Schmidt, Julius, 119
schmidt-cassegrain telescopes, 54, 55, 56, 62, 150
schmidt-newtonian telescopes, 56, 57
schroter effect, 154

Schroter, J.H., 152
Scorpio, 262
secondary mirror, 60
seeing, 102-104
semi-regular variables, 229
setting circles, 90, 91
Shapley, Harlow, 3, 256
sidereal time, 85, 87
Sirius, 14, 103, 213
size, degree measurement, 15
Sky & Telescope, 111, 114, 128, 140, 283
Sky Atlas 2000, 89, 90, 204, 205, 249, 252
Sky Catalogue 2000, 224
sky scanner, 30
Smithsonian Astrophysical Observatory Atlas, 89
sol (Martian day), 159
solar cycles, 280
Solar System, 142-145
Solar System Ephemeris, 186
solar wind, 192
Solis Lacus, 165, 166
spectroscopic binary, 211
spherical aberration, 46, 47, 51, 65, 69, 70
spiral galaxy, 36, 267, 269
sporadic meteors, 206
Stalzer, Robert, 119
star atlas, 86-90, 204, 248
star clusters, 40, 42, 90, 203, 245, 253-262
star hopping, 91, 92, 93, 249
stars, 42
airy disk, 42
internal composition of, 5
magnitude system for, 14
RR Lyrae, 229, 230
RV Tauri, 229
U Geminorum, 232
UV Ceti, 232
variable, 3, 5, 31, 90, 227-245, 317-318
starspots, 228
Stauffer, Mark, 219, 224
Straight Wall, The, 128
strip sketches, 181, 183
Struve, F.G. Wilhelm, 209, 211, 212, 217
Sun, 273-287
dangers of observing, 273
distance to, 9
eclipses of, 287
landscape of, 280-283

report form for, 285
telescopes and filters for, 274-280
wobble of, 283, 284
sunspots, 281, 282, 283, 285
superior conjunction, 143
superior planets, 143, 156
supernovae, 269, 270
suppliers, 325-329
Surveyor, 111, 112
Syrtis Major, 166

T

Taurus, 35, 36, 257
telescopes, 39-65
aperture of, 40
cassegrain reflector, 51, 52
catadioptric, 53-57
collimation and alignment of, 58-63
contrast limits for, 246
dobsonian, 54, 247
double star observation selection of, 213-218, 224-226
focal length and ratio in, 39
image brightness and, 42, 245
light gathering power of, 41
light grasp of, 40-42
limiting magnitudes, 41
magnification of, 43-45
maksutov, 57
Mars observation, selection of, 157-159
newtonian, 51, 52, 146
observations with, 28-37
optical imperfections in, 45-46
planetary observation choices, 145-151
rayleigh and dawes limit of, 43
reflecting, 3, 40, 46, 50-52, 146, 148
refracting, 2, 39, 40, 45-50, 146, 147
resolution, 42
resolving power, 42
rich field, 52, 53, 70, 248
schmidt-cassegrain, 55, 56, 150
schmidt-newtonian, 56, 57
selecting and testing binoculars for, 27-28
selection criteria for, 57
shapes, sizes, specifications, 23-27
solar observations selection, 274-280
star testing, 63-65
testing, 58-65
tri-schiefspiegler, 149